C. H. Calisher
M. C. Horzinek (eds.)

100 Years of Virology

The Birth and Growth
of a Discipline

SpringerWienNewYork

Prof. Dr. Charles H. Calisher
Arthropod-borne and Infectious Diseases Laboratory,
Colorado State University, Fort Collins,
Colorado, U.S.A.

Prof. Dr. M. C. Horzinek
Department of Infectious Diseases and Immunology,
Institute of Virology, State University of Utrecht,
Utrecht, The Netherlands

©1999 Springer-Verlag/Wien
Printed in Austria

Typesetting: Thomson Press (India) Ltd., New Delhi
Printing: A. Holzhausens Nfg., A-1070 Wien
Binding: Fa. Papyrus, A-1100 Wien

Printed on acid-free and chlorine-free bleached paper

SPIN: 10741373

With 75 Figures

CIP data applied for

ISBN 3-211-83384-6 (hard cover) Springer-Verlag Wien New York
ISBN 3-211-83360-9 Archives of Virology [Suppl 15]
(soft cover) Springer-Verlag Wien New York

Preface

Every scientist has key experiences, encounters and quotes that are forever remembered. Writing the editorial to a volume, such as this one on the history of virology, brings them back and puts one's own life in perspective. The visit of Sven Gard, the grand old man of Swedish virology and former Archives editor, is remembered at the Federal Research Institute of Animal Virus Diseases in Tübingen/Germany, where one of us was working at the time as a Ph.D. student. Gard's advice "... you should go into immunology; everything interesting has already been discovered in virology ..." was wrong, at least in its last part. That memorable visit took place in 1965, before reverse transcriptase, polymerase chain reaction, cellular *onc* genes, gene splicing, etc. had been discovered; and before the molecular evolution of viruses was recognized as the source providing virologists with an inexhaustible plethora of subject matter (and relative job security).

History provides not only information for the sake of information but provides perspective on where we are headed. The old adage about redoubling one's efforts when one has lost sight of one's goals never has been more pertinent than today. In this age of molecular wonders it is easy to forget why we are doing this work.

It seems fitting that we have published here selected papers presented at two geographically distant but historically close locations. Amsterdam and Greifswald – or rather Delft and Riems Island – provided the intellectual climate, the seminal insight that led to the birth of a new discipline in microbiology. Its growth in the following years was not steady but occurred in leaps, usually following technological breakthroughs. Many of the early achievements were published in virology's first and most venerable journal, the Archiv für die gesamte Virusforschung. This was founded by Doerr in 1939 and, after the name was changed to Archives of Virology, continued by inspired Editors-in-Chief, the last one being Fred Murphy.

Also with respect to the history of virology, the Archives (a.k.a. "the yellow journal") has a tradition: the Virology Division's News column has run a series of retrospectives, a list of which the reader can find below.

George Santayana said, "Those who cannot remember the past are condemned to repeat it." When one looks at the rediscovery of known viruses (lactic dehydrogenase-elevating virus, for one), this historic imperative provides quite a special flavor. Science is about discovery, so to find something new, one must be aware of what is old. We trust the present volume will convey this message.

We are grateful to Thomas Mettenleiter, who organized the Greifswald meeting in the splendid rococo aula of the University, and to Ab van Kammen and Peter Rottier who did the same at the Royal Academy in Amsterdam; both parties agreed on having selected papers from the meetings published in a joint volume. Special thanks go to the many people who did the real work in producing this issue, the authors of the papers included herein. Their work has been based on the work of others and has served and will serve as a basis for future work by future generations. If this issue provides historical perspective to one young person, we will have succeeded in our task.

References

1. Melnick JL (1991) The beginnings of the International Congresses for Virology. Arch Virol 116: 295–300
2. Bos L (1995) The embryonic beginning of virology: unbiased thinking and dogmatic stagnation. Arch Virol 140: 613–619
3. Horzinek MC (1995) The beginnings of animal virology in Germany. Arch Virol 140: 1157–1162
4. Porterfield JS (1995) The National Institute for Medical Research, Mill Hill. Arch Virol 140: 1329–1336
5. Lindenmann J (1995) The National Institute for Medical Research, Mill Hill: personal recollections from 1956/57. Arch Virol 140: 1687–1691
6. Desmettre Ph (1995) A century of veterinary vaccinology: the Mérieux initiative. Arch Virol 140: 2293–2301
7. Wagner RR (1996) Reminiscences of a virologist wandering in Serendip. Arch Virol 141: 779–788
8. Joklik WK (1996) Famous Institutions in Virology: The Department of Microbiology, Australian National University, and The Laboratory of Cell Biology, National Institute of Allergy and Infectious Diseases. Arch Virol 141: 969–982

Marian C. Horzinek
Charles H. Calisher

Contents

Beijerinck's contribution to the virus concept – an introduction

A. van Kammen

Laboratory of Molecular Biology, Wageningen Agricultural University,
Wageningen, The Netherlands

Summary. The existence of viruses was first recognized when certain pathogens were found to pass through filters that otherwise stop bacteria. Pasteur made such observations in 1887 with the pathogen of rabies, but he thought that the pathogen was a very subtle microbe. In 1886 Adolf Mayer studied the mosaic disease of tobacco plants. He was unable to observe the least trace of a microbe, but still assumed that the pathogen was a bacterium. In 1892 Iwanovsky demonstrated that tobacco mosaic was caused by an agent that passed through bacteria-proof filters but he insisted till the end of his life that the tobacco mosaic virus was a small bacterium. Similar observations were made by Loeffler and Frosch in 1898 on foot-and-mouth disease of cattle. Beijcrinck confirmed the filterability of tobacco mosaic virus but confirmed its properties in more detail and then, in 1898, firmly concluded that tobacco mosaic virus is not a microbe but a contagium vivum fluidum. His idea that a pathogen can be a soluble molecule that proliferates when it is part of the protoplasm of a living cell was revolutionary and new. This new concept has laid the foundation of virus research and directed further studies on the nature of viruses.

*

In 1876, Martinus Willem Beijerinck, then twenty-five years old, was appointed teacher of botany at the Agricultural School in Wageningen, that much later became Wageningen Agricultural University (Fig. 1).

One of his colleagues was Adolf Mayer, a chemist from Heidelberg, Germany, who had come to Wageningen in the same year to teach agricultural chemistry. Mayer's attention was drawn to a serious disease in tobacco, which was, at that time, grown in the region west of Wageningen. The disease caused great losses in yield, and the leaves could not be used for the production of cigars. Beijerinck was first absorbed in continuing his research on plant-galls, which had been the subject of his doctoral dissertation.

Mayer named the disease 'tobacco mosaic'; he demonstrated that it is an infectious disease, and that it can be transmitted to healthy plants by inoculation of sap. He also observed that the infectious agent was inactivated by heating the

Fig. 1. Martinus W. Beijerinck, shortly after he was appointed teacher of botany, in 1876, at the Agricultural High School in Wageningen

leaf juice of infected plants at 80 °C [23]. He concluded that the disease was caused by a bacterium, the infectious form of which he was not able to identify. Beijerinck was very interested in Mayer's experiments on tobacco mosaic and, upon Mayer's request he attempted to identify the responsible microorganism, but failed. He did not attach much value to this lack of success, as he considered himself not a sufficiently trained bacteriologist to solve the problem unambiguously.

In 1885, Beijerinck moved to Delft as he had accepted a position at the Netherlands Yeast and Spirit Works, now grown into Gist-Brocades NV. He became Head of the first industrial research laboratory in the Netherlands. There he developed into a microbiologist in heart and soul, although he continued to do research on plants. Beijerinck became a real microbe hunter, as is illustrated by the many papers on the identification and characterisation of various microorganisms. He had, however, no strong affinity to the technological problems of the Yeast Factory, rather a preference for fundamental academic problems.

In 1895, he acquired a position at the Polytechnical School in Delft, now the Technical University Delft, as a professor of bacteriology, and he was granted a new laboratory and greenhouse facilities (Fig. 2). This should become the cradle of the renowned Dutch School of Microbiologists.

Here Beijerinck took up his studies on tobacco mosaic disease, which he had started in Wageningen. He now demonstrated that the sap of diseased plants was infectious even after filtration through a bacterium-proof porcelain filter candle that retained all visible aerobic bacteria. No microorganisms, neither aerobic nor

Fig. 2. Martinus W. Beijerinck at the age of 45 years, when he had become professor of bacteriology at the Polytechnical School in Delft and resumed his study on the causative agent of tobacco mosaic

anaerobic microbes, could be detected in the infectious filtrate, and the infectious agent could not be cultured in vitro. Moreover, if a drop of juice of diseased plants was put on the surface of a thick agar layer, the contagious principle diffused into the gel, leaving behind all bacteria as well as possible spores. This convinced him that the agent was soluble in water and rather not a microorganism. Furthermore, he showed that the infectious principle is inactivated by heating the juice to 90 °C, thereby excluding the possibility of dealing with spores.

The agent actually multiplied in living tissues of infected plants. The rapidly growing young leaves of tobacco plants were particularly affected and showed severe disturbances in development. In addition, the material could be stored for months without losing its infectious properties, even in soil, and it could be precipitated with alcohol without loss of infectivity. After many painstaking experiments, unable to detect bacteria by any accepted bacteriological technique, Beijerinck was brought to conclude that tobacco mosaic disease is caused not by a bacterium, but by a *contagium vivum fluidum*. That resulted in the paper he presented to the Royal Netherlands Academy of Sciences on November 27, 1898, which is commemorated this year [3, 4]. A more detailed description of the experiments is given in his publication of 1900 in the *Archives Neërlandaises des Sciences Exactes et Naturelles*. A biography of Beijerinck was published by Iterson et al. [17].

In 1892, Iwanovsky, in Russia, had already demonstrated that the agent causing tobacco mosaic passed through bacteria-proof filters. Loeffler and Frosch [20] made similar observations on the infectious principle of foot- and mouth disease; however, these researchers concluded and maintained that a small microbe was involved. In doing so, they conformed themselves to the authority of Pasteur [24], who had stated that 'virulent affections are caused by small microscopic beings, which are called microbes... The microbe of rabies has not been isolated as yet, but judging by analogy we must believe in its existence. . . to resume: every virus is a microbe.'

The discovery and definition by Beijerinck of a category of infectious agents differing from all microorganisms was therefore new and revolutionary – a new concept. The idea that tobacco mosaic virus (TMV) is a *contagium vivum fluidum*, is molecular and soluble, cut across the idea of a microorganism. The *contagium vivum fluidum* defined a self-reproducing, subcellular entity.

Beijerinck's ideas met with strong opposition and were not readily accepted. In the next decade, however, more self-reproducing pathogenic agents were found that shared the properties of being filterable, invisible by microscopy and impossible to culture in vitro. Amongst them were the causative agents of several other plant diseases and of animal diseases such as measles, poliomyelitis, rabies, yellow fever and smallpox.

Beijerinck's concept obtained support from the discovery of bacteriophages by Twort in 1915 [29]; subsequently D'Hérelle [15] showed that bacteriophages can destroy bacteria but also require bacteria for their multiplication. D'Hérelle gave full credit to Beijerinck as being the first scientist to declare that there can be infectious agents replicating at the expense of the living cell, and which are themselves non-cellular and much smaller than cells. Such agents might be proteins like albumin, or enzymes.

Indeed the failure of microscopic and cultural methods to reveal any cause of virus diseases as established by Beijerinck and others had produced various speculations on the nature of the *contagium vivum fluidum*, and often proteins and enzymes were indicated. Beijerinck referred to these hypotheses in his paper on *The Enzyme Theory of Heredity* [6] where he suggested the application of methods that were being developed in protein chemistry. Thus, Mulvania [22] found that TMV could be precipitated with protein precipitants without loss of infectivity.

In 1926, Summer obtained the first enzyme, urease, in pure form and produced crystals from pure urease – a milestone in protein chemistry. Another major development was the demonstration that the local lesions produced by TMV on the leaves of some host plants could be used for a quantitative assay, similar to the plaque test for bacteriophages.

Inspired by the success of purification of an enzyme, Wendell M. Stanley attempted to purify and isolate TMV using precipitation with ammonium sulphate. In 1935, he obtained crystals of TMV from the juice of infected tobacco plants and concluded that TMV was to be regarded as an autocatalytic protein that required the presence of living cells for multiplication [27, 28].

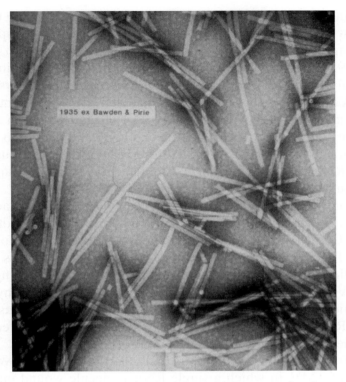

Fig. 3. Electron micrograph of negatively stained tobacco mosaic virus particles from a preparation purified by Bawden and Pirie, in 1935, at Rothamsted Experimental Station, England. The electron micrograph was made fifty years later, in the eighties (courtesy by Dr. T. M. A. Wilson, Dundee, Scotland)

One year later, Bawden and others [1, 2] in England showed that TMV is not really a pure protein, but contains about 5% ribonucleic acid (RNA) (Fig. 3). Two years earlier, in 1934, Schlesinger had shown that bacteriophages contain protein and deoxyribonucleic acid (DNA). Nucleic acid had made its entrance in virus research.

These findings represented major discoveries that strongly appealed to the imagination. How could an agent like TMV reproduce if its chemical composition was so simple? It became a major goal to elucidate the structure of viruses in order to learn how they work. Viruses received full attention by biophysicists, biochemists and crystallographers. Part of the studies of viruses were molecular studies, actually leading to molecular biology.

The first indications of the shape of TMV particles came from the observation that dilute solutions of TMV showed the phenomenon of anisotropy of flow. Bawden et al. [2] used a goldfish swimming in a TMV solution to demonstrate this phenomenon and concluded that the virus particles were probably rod-shaped.

The real disclosure of the shape and form of virus particles was brought about by the development of the electron microscope in the late thirties, which made direct visualisation of the hitherto invisible particles possible. It revealed that TMV particles are indeed rod-shaped [19] and as we know now, are 300 nm in length and 18 nm in width. Many other viruses are spherical. The more complex structure of bacteriophages, with their heads and tails, was first visualised in 1940 [25].

Then in 1952, Hershey and Chase discovered that, upon infection, the phage DNA is the only, or at least the principal phage component that enters the host cell while the bulk of the phage protein remains outside. This experiment showed that DNA is the carrier of the genetic programme for phage replication and provides for the genetic continuity of the phage. Later, Gierer and Schramm [13, 14] and Fraenkel-Conrat [9] proved that the RNA from TMV is the infectious component and that viral RNA, devoid of protein, can initiate virus replication upon infection of tobacco plants.

In the meanwhile, X-ray diffraction studies of TMV crystals had shown that virus particles are assembled from a large number of identical protein subunits [8]. Fraenkel-Conrat and Williams [10] then found that in a solution containing a mixture of TMV RNA and disassembled protein subunits, virus particles reconstitute to their original structure and regain full infectivity. Using RNA from different TMV-strains, Fraenkel-Conrat and Singer [11] prepared hybrid viruses by reconstitution and showed that, after infection, the probing virus was always composed of RNA and protein corresponding to the RNA in the infecting hybrid. This clearly demonstrated RNA as the genome of the virus.

X-ray diffraction studies by Franklin and others [12] finally revealed the entire structure of TMV, in which a single RNA molecule wound into a helix, is surrounded by radially arranged protein subunits. By then, the particle of the *contagium vivum fluidum* responsible for tobacco mosaic disease had been characterised in all detail.

Through the years, a large and ever increasing number of animal, plant and bacterial viruses has been identified. In the middle of this century, almost fifty years after Beijerinck's discovery, virus research had gradually developed into virology, a branch of biological science constituting a distinct body of knowledge and methodology, with its own genetics and generalisations that formed a firm basis for further research. There was a general agreement that viruses are entities whose genomes are elements of nucleic acid, either DNA or RNA. They replicate inside living cells, they use the cells' metabolic and protein synthetic machinery and direct the synthesis of specialised elements that can transfer the viral genome to other cells. Viruses neither grow nor divide, so they are no organisms. They depend upon host organisms for their reproduction. During replication, they become part of the infected cell, but they have their own genetic programme.

To obtain an overview of the large variety of viruses and their biological properties, and to reveal their possible evolutionary relations it became essential to classify viruses in meaningful categories [21]. By then, Beijerinck had been dead for more than 25 years. He died January 1, 1931 and did not live to see the purification and crystallisation of TMV. At his retirement from the University of Delft in 1921, he ended his farewell address by proclaiming '. . .how happy are those who are now beginning. . .' Might he have had some notion of how exciting virus research would become and how important for biology?

References

1. Bawden FC, Pirie NW (1937) The isolation and some properties of liquid crystalline substances from solanaceous plants infected with three strains of tobacco mosaic virus. Proc R Soc London Ser B 123: 274–320
2. Bawden FC, Pirie NW, Bernal JD, Fankuchen I (1936) Liquid crystalline substances from virus-infected plants. Nature 138: 1 051–1 052
3. Beijerinck MW (1898a) Over een Contagium vivum fluidum als oorzaak van de vlekziekte der tabaksbladen. Versl Gew Verg Wis en Natuurk Afd., 26 Nov. 1898. Kon Akad Wetensch Amsterdam VII: 229–235
4. Beijerinck MW (1898b) Über ein Contagium vivum fluidum als Ursache der Flekkenkrankheit der Tabaksblätter. Verh Kon Akad Wetensch VI: 3–21 [English translation (1942) Concerning a contagium vivum fluidum as cause of the spot disease of tobacco leaves. Phytopathol Classics 7: 33–52]
5. Beijerinck MW (1900) De l'existence d'un principe contagieux vivant fluide, agent de la nielle des feuilles de tabac. Arch Néerl Sci Ex Natur, Haarlem, Ser II: 3, 164–186
6. Beijerinck MW (1917) The enzyme theory of heredity. Proc Section of Sciences, Kon Akad Wetensch 19: 1 275–1 289
7. Beijerinck MW (1921) Verzamelde Geschriften van M.W. Beijerinck. M Nijhoff, The Hague [in five volumes; a sixth volume published in 1940 contains a biography and papers by Beijerinck, which appeared after 1920]
8. Bernal JD, Fankuchen I (1941) X-ray and crystallographic studies of plant virus preparations. J Gen Physiol 25; part 1: 111–165; part 2: 120–146; part 3: 147–165
9. Fraenkel-Conrat H (1956) The role of nucleic acid in the reconstitution of active tobacco mosaic virus. J Am Chem Soc 78: 882– 883
10. Fraenkel-Conrat H, Williams RC (1955) Reconstitution of active tobacco mosaic virus from its inactive protein and nucleic acid components. Proc Natl Acad Sci US 41: 690–698
11. Fraenkel-Conrat H, Singer B (1957) Virus reconstitution. II. Combination of protein and nucleic acid from different strains. Biochem Biophys Acta 24: 540–548
12. Franklin RE, Klug A, Holmes KC (1957) X-ray diffraction studies of the structure and morphology of tobacco mosaic virus. Ciba Found Symp: The nature of viruses. Churchill, London, pp 39–55
13. Gierer A, Schramm G (1956a) Die Infektiosität der Nukleinsäure aus Tabaksmosaikvirus. Z Naturforsch 116: 138–142
14. Gierer A, Schramm G (1956b) Infectivity of ribonucleic acid from tobacco mosaic virus. Nature 177: 702
15. D'Hérelle F (1921) Le bacteriophage. Masson, Paris
16. Hershey AD, Chase M (1952) Independent functions of viral protein and nucleic acid in growth of bacteriophage. J Gen Physiol 36: 39–56
17. Iterson Jr G van, Dooren de Jong LE den, Kluyver AJ (1940) Martinus Willem Beijerinck. His life and his work. M Nijhoff, The Hague [reprinted 1983 by Science Tech, Madison]
18. Ivanovsky D (1892) Über die Mosaikkrankheit der Tabakspflanze. St Petersb Acad Imp Sci Bull 35: 67–70 [English translation (1942) Concerning the mosaic disease of tobacco. Phytopathol Classics 7: 25–30]
19. Kausche GA, Pfankuch E, Ruska A (1939) Die Sichtbarmachung von pflanzlichen Virus im Übermikroskop. Naturwissenschaften 27: 292–299
20. Loeffler F, Frosch P (1898) Berichte (I–III) der Kommission zur Erforschung der Maulund Klauenseuche bei dem Institut für Infektionskrankheiten in Berlin. Zbl Bakt Parasitenkr I 23: 371–391

21. Lwoff A, Horne RW, Tournier P (1962) A system of viruses. Cold Spring Harbor Symp Quant Biology. Basic Mech Anim Virus Biol 27: 51–55
22. Mulvania M (1926) Studies on the nature of the virus of tobacco mosaic. Phytopathology 16: 853–871
23. Mayer AE (1886) Über die Mosaikkrankheit des Tabaks. Landw Versuchsstation 32: 451–467 [English translation (1942) Concerning the mosaic disease of tobacco. Phytopathol Classics 7: 11– 24]
24. Pasteur L (1890) La rage. Lecture 65: 449–465
25. Pfankuch E, Kausche GA (1940) Isolierung und übermikroskopische Abbildung eines Bakteriophagen. Naturwissenschaften 28: 46
26. Schlesinger M (1934) The Feulgen reaction of the bacteriophage substance. Nature 138: 508–509
27. Stanley WM (1935) Isolation of a crystalline protein possessing the properties of tobacco mosaic virus. Science 81: 644–645
28. Stanley WM (1936) Chemical studies on the virus of tobacco mosaic. VI. Isolation from diseased Turkish tobacco plants of a crystalline protein possessing the properties of tobacco mosaic virus. Phytopathology 26: 305–320
29. Twort FW (1915) An investigation on the nature of ultra-microscopic viruses. Lancet ii: 1 241–1 242

Authors' address: Dr. A. van Kammen, Laboratory of Molecular Biology, Wageningen Agricultural University, Dreijenlaan 3, 5703 HA Wageningen, The Netherlands.

The Prussian State and microbiological research – Friedrich Loeffler and his approach to the "invisible" virus

H.-P. Schmiedebach

Institut für Geschichte der Medizin, Greifswald, Federal Republic of Germany

Summary. When Loeffler took his first steps in the newly-emerging field of virology, the aim and the methods of his research activities were influenced by two different issues: 1) Loeffler was rooted in the scientific paradigm of bacteriology, but during the progress of his research on foot-and-mouth-disease (FMD) he recognized that the classical techniques derived from bacteriology were useless in identifying the agent of this disease. Thus he focussed on the properties of the pathogen and – though he could not find a method in order to visualize the 'virus' – he tried to develop a vaccine against the disease. 2) The Prussian Government was highly interested in effectively combatting FMD. In 1897 Loeffler was appointed by the Ministry of Cultural Affairs to the newly-established commission for exploring that disease. The agricultural lobbies and the public pursued the activities of the commission with a mixture of hope and serious scepticism and demanded convincing results. These circumstances caused a considerable degree of political pressure on Loeffler, pressure which determined that his research activities would take a pragmatic approach, that he would avoid sophisticated reflections and trials on the nature of the 'virus', and that his research strategies would have as a goal the development of an effective immunization.

Introduction

In 1897, when Loeffler was appointed to the commission that was to explore FMD he already was a highly respected bacteriologist. As one of the first collaborators with Robert Koch, Loeffler had worked with him for about eight years (Fig. 1). During this time, he became familiar with the methods of bacteriological research of that era, but in 1888, he accepted the newly established Chair for Hygiene at the medical faculty in Greifswald. In the years before his appointment to Greifswald he had been able to make some outstanding discoveries in the field of bacteriology; for example, in 1884 he identified organism causing diphtheria and about five years later he isolated the toxin produced by this organism [29–31].

In this paper I discuss his merits concerning the establishment of virology. Loeffler is often named the "father of virology", mostly in German or German-

Fig. 1. Friedrich Loeffler and Robert Koch (about 1886) [source: Unger H (1936) Robert Koch. Roman eines großen Lebens, 6. Aufl. Verlag der deuschen Ärzteschaft, Berlin]

speaking periodicals, while the Anglo-American journals primarily apply this title to Dimitri Ivanovski [17] or to Martinus Willem Beijerinck [13]. Ivanovski is generally given credit for first recognizing an entity that was filterable, sub-microscopic in size, and that might be the cause of a disease [28, 37]. In 1892 he presented a paper before the Academy of Sciences of St. Petersburg in which he stated that the sap of leaves infected with tobacco mosaic disease retained its infectious properties after filtration through filter candles [16]. However, this discovery produced several questions concerned with the nature of the filterable infectious agent [14]. What did Beijerinck mean when he spoke of a *contagium vivum fluidum* (contagious living fluid) [4]? Was the agent hypothesized as liquid and soluble or was it as particulate? Was the ability to multiply a kind of self-reproduction of a living organism or was it more a product of metabolic activity of the host-organism? The evaluation of these questions was strictly connected to the emergence of virology.

In this paper I deal with two aspects of Loeffler's contribution to virology and his exploration of FMD: the first issue emphasizes the research goals and the experimental settings. The second focuses on the contextual background in which the experimental strategies of Loeffler, and his collaborators Frosch and Uhlenhuth, were implemented. Both aspects are interwoven to some extent and constitute a reciprocal relationship. The contextual conditions embrace political demands, practical purposes of the research project as well as the pressure and the power of public opinion with regard to the expected success of the research activities. The experimental work should be evaluated with respect to the trials they performed, the devices they used and the problems they tried to overcome.

The published records and articles written either by the research commission, which Loeffler headed, or by himself show some epistemological uncertainties. Most of the authors who have described Loeffler's merits ignored such passages in his articles and did not discuss these interesting items [1, 3, 6, 7, 8, 10, 15, 33–36]; but even Claude Bernard, one of the founders of the experimental method in physiology, stated in 1865 that when the scientific object is absolutely dark and unexplored, the physiologist is allowed to act haphazardly and to undertake something that Bernard likened to "fish in troubled waters" [5]. To emphasise such dark aspects of a research process leads us to the most interesting issues; if we discuss these questions we can discover the different factors that determine the proceedings and the results of those activities. The thesis that emerges from this and that I discuss is as follows:

The research programme concerning FMD, which was initiated by political authorities, was primarily aimed at the development of an effective vaccine but led, as a side-effect, to the virus itself. Nonetheless, it was not possible to find a way to visualize the supposed virus and thus Koch's postulates could not be completely employed. Loeffler emphasized the practical side of his research activities, which to some extent de-emphasised the theoretical and scientific requirements. The political, practical and public context determined the experimental strategy and the focal points of his research.

The experimental settings

I refer to the experimental setting, or to the history of the discovery, as far as it is concerned with the establishment of virology. The first fact we take into consideration is that the term "virus" was used in discussions in papers by bacteriologists long before the end of the 19th century. Even in Antiquity and in the Middle Ages the term "virus" denoted a venomous agent. In 1844 a well-known German medical encyclopedia gave several definitions, ranging from "poison", to "miasma" and to "contagium" [34]. In the first records of Loeffler and Frosch in the year 1897, this notion is mentioned several times. Of course this term did not then embrace the connotations of the modern word "virus". They used this word in the sense of an agent that causes a disease ("Krankheitsstoff") [25]. About one year later Loeffler added the denotation of an agent with the ability to multiply [18].

Loeffler and Frosch adhered to the paradigm of bacteriology and all their research activities moved within this field. They were looking for a bacterium or a bacterium-like germ and they used the approved methods of bacteriology, for example the culture media [25, 27]. The result of these trials was a definite one: a bacterium could not be found and certain cultures contained organism that were contaminants and which were easy to identify.

Fig. 2. Loeffler in his laboratory [source: Unger H (1941) Unvergängliches Erbe. Das Lebenswerk Emil von Behrings. Gerhard Stalling Verlagsbuchhandlung, Oldenburg Berlin]

In a second step Loeffler and Frosch focussed on some of the very important properties of the germ they were looking for, properties relevant to the developement of a strategy to be used against the infectious agent. Therefore, they formulated seven research issues: 1) transmitting the disease to various species of animals; 2) mode of infection; 3) the infectious material itself; 4) the duration of activity of the "virus", 5) methods for destruction of the pathogen; 6) the development of immunity; and 7) the possibility of vaccination [25]. In all, they established standard methods that could be used to study many viruses.

During half a year of experimentation on these problems they emphasized the question as to whether animals that recovered from the disease were then protected by an immunity against a second infection. They spoke about the dangers to the agricultural production due to this disease. Not only the large number of dead animals, but also losses of milk, meat, capacity for work and the negative influence on cattle-breeding were mentioned in this regard.

In addition to this problem, there were contradictory opinions about the question of immunity. Some veterinary authorities denied the possibility of immunizing against FMD. However, certain observations made by the commission showed that some of the animals did indeed mount immunity. Nonetheless, this was not a consistent occurrence and there was no knowledge about the duration of such protection.

Later they tried to determine a useful immunization procedure. They performed some trials with different kinds of lymph, with the blood of animals that were shown to be immune and with a mixture of lymph and blood of these protected animals. Through these trials they discovered that blood from animals that had recovered from FMD could protect other animals [25].

With regard to the question of immunization the commission conducted other experiments. They injected blood from newly affected animals with and without blood of immunized individuals. In another experiment, they removed lymph, which was cleaned of any corpuscular elements by a filtration process, diluted it with 39 parts of water, mixed in bacteria as a marker, and filtered this twice through a Berkefeld filter. The filtration was considered as successful when there were no bacteria in the filtered substance, proving that all bacterial and corpuscular elements were held back in the filter.

This kind of filtration had been used since the early 1870s in order to remove microorganisms from fluids. Loeffler and Frosch wanted to determine whether there were soluble agents within this lymph that could provide immunity to FMD. The aim of these injections was strictly connected to the immunization. The investigators were surprised when they saw that all cows treated with the filtered lymph became as ill as the control cows that had received injections of unfiltered material.

Reflecting on this phenomenon they considered two possibilities: First, that the filtered lymph, which did not contain any bacteriological elements, contained a soluble, extremely efficient toxin and second, that the invisible germs causing FMD were of such a small size that they could pass the pores of a filter able to hold back all known bacteria. Loeffler and Frosch reported this in their third record, from January 1898 [25]. As mentioned, Loeffler was experienced with both alternatives. In the past he had discovered some germs and in 1889 he had been able to isolate the toxin of the organism causing diphtheria. So Loeffler now had to construct an experimental strategy to exclude or to prove one of the alternatives.

In a first step he compared the toxic efficacy of lymph from cattle with FMD with that of tetanus-toxin. In an arithmetical example he came to the conclusion that the toxic efficacy of such lymph would be much larger than that of the very highly effective toxin of tentanus. He then referred to an observation he had made with a pig that became infected with FMD after injection with a diluted sample of lymph taken from a blister of a cow that had become ill after being injected with diluted filtered lymph. Because in this example the toxin had been diluted twice, the supposed toxin efficacy would be extremely high. Loeffler commented on the result of his calculations with the words: "Such a toxic efficacy would be simply incredible!" [25, p.100] with this rhetoric he arrived at the second step.

Once again he presented a mathematical example combined with a simple experimental observation. The starting-point was an assumption. He assumed that the toxin of FMD that would be totally, or to a very large extent, excreted at those places of the body-surface where the blisters emerged. He felt this assumption was legitimate because it was not possible to prove the existence of a toxin in organs of a deceased animal. Moreover, he estimated that the whole content of all blisters would be 5 ml. With only 1/30 ml of filtered lymph it was possible to infect an animal. Accordingly the original amount of toxin was now diluted to 1:150. Only 1/50 ml of this diluted lymph was enough to infect a 30 kg pig, dilution of 1/7500 of the original fluid. Again he compared his results with the toxic efficacy of tetanus toxin and related it to a gram of blood: it would result in a toxic value of the FMD-toxin per gram of pig blood of 1:7,500,000,000. Also this second example seemed to prove a high efficacy of the toxin. Loeffler then presented his conclusion – based on his assumptions and on these calculations-which he formulated very cautiously: "Thus we cannot reject the assumption that the effect of the filtered lymph is not caused by a soluble substance, but rather by a germ with the ability to multiply" [25, p. 100]. („Es läßt sich deshalb die Annahme nicht von der Hand weisen, daß es sich bei den Wirkungen der Filtrate nicht um die Wirkungen eines gelösten Stoffes handelt, sondern um die Wirkung vermehrungsfähiger Erreger").

These calculations, based on a mechanical view without consideration of possible metabolic activities, brought him to the conviction that there existed a germ of very small size that could pass through common filters. Nonetheless, he could not present scientifically derived evidence for the existence of that small germ. Loeffler was fully aware of this epistemological dilemma. In his record he added an explanation for the impossibility of seeing this very tiny germ. He referred to correspondence with Professor Abbé in Jena who was an authority of the highest reputation regarding microscopic techniques. Loeffler discussed with him the limitations of microscope performance. If the supposed germ of FMD had a size of only about 0.1 μm, even the best immersion techniques of that day could not made this virus visible. According to Loeffler, this would be best explanation for the fruitless attempts to discover the germ by microscopic investigation. Although this was a very pessimistic view, he tried to turn the tables and offered a perspective concerning the possible discovery of a large number of other germs that could not be identified at that time. In connection with the necessity of future studies on that problem, the commission also requested for the grant of new financial support from the government [25].

The commission could not identify the supposed microbe by microscope nor was it possible to make it visible by any other methods. However, there were some scientific requirements to be fulfilled in order to accept a supposed germ as the causative agent for a disease. In 1903 Loeffler himself wrote on the occasion of Koch's 60th birthday about the scientific foundations of the newly emerged discipline of bacteriology and declared the development of "Koch's Postulates" as a decisive attainment. While Koch had mentioned four postulates in 1890 [12], Loeffler referred only to three:

1. "Constant evidence of the concerned organism in all cases of the disease;
2. isolation of the pathogen in a pure culture that had to be cleaned of all corpuscular elements of the sick individual;
3. generation (Wiedererzeugung) of the disease anew by reliable pure cultures." [20]

I will not discuss the differences from Koch's original formulation but we must evaluate whether Loeffler himself undertook any steps to employ these

Fig. 3. Shibasaturo Kitasato in the Institute of Hygiene in Berlin (1899) [source: Collected Papers of Shibasaturo Kitasato (1977). Kitasato Institute and University, Tokyo]

three postulates in the case of FMD and which, if any, he employed. As we have already seen, two very successful methods, microscopy and culture, failed with regard to the identification of the microbe. So Loeffler and his collaborators had to figure out some other methods. As already reported in their first record, the commission focussed on the biological or biochemical properties of the causative agent, properties such as duration of activity, resistance to high temperatures, destruction of the supposed "virus", etc. The published records contain some information on these properties, sometimes with slight modifications from one record to another.

Moving on in this field, the terms "invisible" and "filterable" emerged [24]. Quality of filtration was determined by the quality of the devices and the conditions of techniques such as pore size of the filter, adsorption properties, and filtration pressure. So the category "filterable" became primarily an experimental definition. Moreover, this experimentally defined category did not always apply, even for the supposed pathogenic virus of FMD. In their record from 12 August 1898 Loeffler reported that diluted lymph, which had been squeezed several times through the very small pores of a Kitasato-filter (see Fig. 5) lost its pathogenic ability. Loeffler concluded that the pathogenic agent of FMD could not permeate these very narrow pores; thus it must be of corpuscular character [18].

The agent was sometimes filterable, sometimes not. This experimental definition was dependent on the choice of the filters and of the filtration technique, but there is no hint that Loeffler or one of his collaborators understood the significance

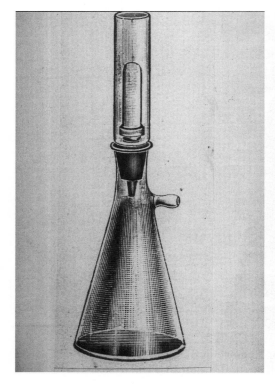 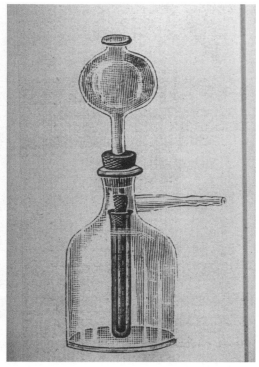

Fig. 4. Berkefeld-Filter [source: Catalogue of   **Fig. 5.** Kitasato-Filter [source: Catalogue of
the firm "Pohl", Den Haag (without year)]   the firm "Pohl", Den Haag (without year)]

of this observation and performed additional experiments to find out other ways
to characterise the supposed virus.

In 1907 Loeffler published an article dealing with new methods of quick
colouring microorganisms. He wrote that the during previous years he had tried
to identify the pathogen of FMD by a special mode of colouring but that all these
trials had failed. In addition, his newly developed procedure based on two highly
effective methods, the malachite colouring and the Giemsa staining method, could
not make the virus visible [27]. So some essential aspects of "Koch's Postulates"
could not be employed in Loeffler's research activities.

Nonetheless, in one respect Loeffler was successful. In 1903, he told the
scientific community that he had found a culture medium for FMD virus, the
bodies of piglets. In his attempts to develop an effective serum against FMD, he
needed lymph with as constant and as high a virulence as possible. The little pigs
of the Yorkshire race proved to be the most suitable animals for this purpose.
In order to continue the cultivation of the virus he injected a certain quantity of
lymph every 5 to 6 days. Normally after 2 to 3 days, the pigs became sick and,
after having lymph removed from the blisters that developed, a protective serum
was injected so that the losses were limited [19]. This success was more a result
of the immunization experiments than a product of a well-calculated hunt for an
adequate culture medium for FMD virus. Despite this success, one hardly can say

that all Koch's postulates were fulfilled. Loeffler was aware of these difficulties and though he did not often give notice about his unsuccessful trials, he obviously could not claim a breakthrough concerning the identification of FMD virus. Owing to this deficiency, in the first decade of the 20th century a discussion about the nature of the newly described "virus" arose. This controversial dispute took place to a large extent in the pages of the "Centralblatt für Bakteriologie, Parasitenkunde und Infektionskrankheiten" and revolved at that time primarily around tobacco mosaic virus [11, 32, 38]. The two protagonists were Beijerinck and Ivanovski. The central item they dealt with was a question of the nature of the virus: was it a living agent or an inert chemical product [14, 37]? It is striking that Loeffler did not contribute to this discussion with articles referring to his own research results. Only in an article from 1906 did he mention in one phrase the "smallest plants", the so-called bacteria [22]; however, he gave only notice of this opinion, without any argumentation or reflection. Even if we concede that these questions were only academic and that convincing evidence based on experimental research was lacking for either approach, it is surprising that Loeffler, as one of the fathers of virology, did not take part in this discussion and did not question the contradictory results with regard to the filtration results.

Political context and public pressure

In discussing the politics of this drama, I refer to what I have introduced as the contextual conditions of the research project on FMD. Two Prussian Ministries played a crucial role concerning financial support. The Prussian Ministry of Cultural Affairs from 1897 to 1907 and, from 1909 onward, the Prussian Ministry of Agricultural Affairs. The establishment of a research commission in 1897 was intended to provide the solution to this practical problem in the field of livestock-breeding and production with the help of the newly established discipline of bacteriology [7].

Losses in the agricultural production caused by FMD were extremely large; at the end of the 19th century they amounted to 100 million marks a year. The appointment of Loeffler to the Chair of Hygiene at Greifswald University in the year 1888 was largely a political decision by the Ministry of Cultural Affairs and not the result of an academic desire of the faculty. The list of names of the proposed scholars for this appointment embraced four persons. The first position was given to Gustav Wolfhügel, who had studied chemistry and medicine and was qualified in hygiene by Max von Pettenkofer in Munich. The second person was Ernst Salkowski, who had worked in Tübingen and later for Rudolf Virchow in Berlin, becoming head of the chemistry department at Virchow's Institute for Pathology. The third candidate was Friedrich Renk from the Berliner Reichsgesundheitsamt. The fourth person was Loeffler, who was only able to gain a place on that list because the Dean's vote for Loeffler was double-counted; thus a majority of one voice ensured position number four for Loeffler. This was a very uncommon procedure. The faculty wrote to the Minister that in the opinion of the professors, Loeffler was indeed qualified in bacteriology but not sufficiently in chemistry,

Fig. 6. Page of the "Berliner Abendpost" from 24 April 1911 with the article that reproached Loeffler about not having discovered the pathogen causing FMD ("Die Tierseuchen-Kalamität")

which was seen as a main pillar of hygiene. Therefore, the faculty preferred an appointment of one of the other candidates [31]. Nonetheless, the Minister was convinced of the high reputation of bacteriology and its relevance to hygiene, and Loeffler was allowed to take up his new position at Greifswald University.

The beginning of Loeffler's research work on FMD was promoted by an activity of the Prussian Government. In 1897, the Prussian Ministry of Cultural Affairs established a "commission for the investigation of the FMD at the Institute for Infectious Diseases in Berlin." The collaborators of Loeffler were, first, the veterinarian Paul Frosch and, later, Paul Uhlenhuth. This commission was obliged to submit regular reports to the Ministry of Cultural Affairs.

The initiative for this research programme, as well as the financial support, were results of political interests of the state. Million mark losses caused by FMD each year in the agricultural sector, especially in the field of milk production and cattle breeding, stimulated the activity of the parliament and paved the way for state support. Unfortunately, this support made Loeffler dependent upon political trend.

However, the Ministry did not restrict itself to these basic activities; it built a network that was intended to provide information necessary to collect the needed material. The Ministry required several local authorities of rural communities around Berlin to send information about new out-breaks of the disease immediately by telegraph to the Berlin Institute for infectious diseases; district veterinarians (Kreistierärzte) particularly supported the commission. Thus, the receipt of fresh lymph samples from newly erupted blisters of cows was guaranteed. In their first collection, Loeffler and Frosch referred to the contents of blisters of 12 animals from four places.

Later, when they had performed successful laboratory trials and found effective sera, they wanted to scrutinize their results under practical conditions in the countryside. They had to wait for such an opportunity and, when informed about a new epidemic, they travelled to the affected farms. In their records they tell us about a large scale of "considerable difficulties" (erhebliche Schwierigkeiten): because of very wild animals, for example, but serious injuries did not happen [25]. Loeffler and Frosch had to receive the permission of the proprietor of the concerned cattle herds before they could vaccinate the animals. Thus the laboratory work was integrated not only into an information network, it was also dependent on farmers in the countryside. All these conditions of their research work underline the practical purpose of the research programmes.

In 1907, Loeffler was confronted with numerous difficulties. Since 1902, he had been using a farm in Greifswald for his trials. In 1907, the Prussian Minister of Agricultural Affairs accused Loeffler of being responsible for the dissemination of FMD in the region of Greifswald. Loeffler's experiments at this farm were seen as the main cause for the spread of the disease.

Supported by agricultural associations of farmers of Pomerania, the Minister demanded the suspension of all experiments at the farm as well as the university institute. Loeffler had to stop his activities and the farm was rented to a farmer and a master carpenter [7]. After an intermission of two years, he resumed his research activities in another place: the newly purchased and equipped island of Riems, which provided an almost ideal location for his work. The danger of disseminating the disease was minimized because it was an island.

Two years later Loeffler had again to endure great public pressure. The success of his vaccination experiments was denied by two important professors: Professor Schmalz, head of the Berlin Veterinary Medical School and Professor Casper from Breslau University. They both strongly criticized Loeffler's research activities. Their opinions were published in an article in a newspaper in April 1911 (see

Fig. 6). The article attacked both Loeffler and the Ministry of Agricultural Affaris. The unknown author blamed the Minister for having given too much credit to Loeffler's immunization methods. Because the Minister had eased restrictive measures, the epidemic spread, infesting about 11,000 farms. Loeffler is said to have announced newly developed sera each year but none showed any effect. He was even reproached because he had not discovered the etiologic agent of FMD [2].

This was motivated by differences between two concepts of combating the disease: Schmalz and Caspar preferred veterinary administrative restriction, such measures seen as being able to be lifted if a vaccine would be developed. With respect to the uncertain results of veterinary research, the protagonists of veterinary policy measures did not agree with any easing of such restrictions. Not only funding quandaries but also the politically-motivated purpose of the research project as well as the public pressure were contextual factors influencing the specifies of the research.

Closing remarks

1. As mentioned, Loeffler was appointed by the Ministry of Cultural Affairs to a commission for studying FMD. The entire project was dedicated to a very practical end: combating the heavy agricultural losses caused by FMD.
2. The experimental setting, a laboratory as a center of a widespread network of farms and stables in Pomerania where trials could be performed in cattle under rural conditions, underlined this practical reference to cattle-breeding and agricultural production.
3. In 1899 Loeffler was appointed as an extraordinary member to the "Kaiserliches Gesundheitsamt". He was obliged to observe the development of all aspects of public health in his district. With this appointment, he became someone like a public health officer as controller and advisor in public health affairs.
4. From the beginning of his research activities on FMD, he and his collaborators prioritized the development of a vaccine against the disease. The majority of his statements, and publications dealt with problems of immunization [19, 21, 23, 26]. His first trials in that field had already given some reason for an optimistic assessment and for further study. Attempts to identify the etiologic agent were subsumed to the loftier goal of finding a method for adequate immuniztion. Loeffler hoped that the identification of the virus could facilitate and lower costs of production of an effective vaccine.
5. The agricultural lobbies and the public, especially the rural population, pursued the activities of the commission with a mixture of hope and scepticism. From 1907 on, Loeffler was forced to interrupt his research activities because of the resistance of agricultural associations, but in 1909, the Ministry of Agricultural Affairs again granted financial support. However, now the pressure on Loeffler increased. The Ministry of Agricultural Affairs had to decide which protective measures should be performed in case of an epidemic. These measures embraced a grand scale, ranging from temporary segregation of affected

farms to the slaughter of sick animals. Loeffler was an important advisor to the Ministry and so he had to take on a considerable degree of responsibility. If he was successful with his research on vaccination, the toughness and duration of protective measures ordered by the political authorities could be minimized.

Dependent on the success of medical innovations was a dynamic relationship between governmental regulations and veterinary medicine. The more effective a vaccine was, the less rigorous veterinary policy measures would become. Without an effective vaccine, the extensive veterinary administration concept would remain dominant. There were certain contradictions between these two approaches, although in 1909 Loeffler tried to construct a more complementary relationship. He was fully aware that an effective protection, i.e. active immunity, was attained only five weeks after immunization. Passive immunity induced by a serum became effective at once but it lasted only for two weeks. These facts demonstrate distinct limitations for a dominance of veterinary prophylaxis over administration. Therefore, he stated that a serum vaccination would not be the only measure against the epidemic and that a real effect would be caused when passive immunity was applied in combination with rigorous veterinary administration measures [23]. According to the veterinary authorities at that time, preventive measures had failed, so strict veterinary regulations seemed to be the only way to combat FMD successfully.

One can imagine which political pressure determined Loeffler's research activities. He always accepted the very practical purpose of the FMD-research project and he submitted his research strategies to the goal of developing an effective method of immunization. Therefore he used the methods he had become familiar with during his time as Koch's collaborator, but his overall activity was concerned with the exploration of FMD, giving special attention to the practical end. In particular the Ministry of Agricultural Affairs demanded effective results for its continued financial support and Loeffler did perform the first successful trials of immunization. This led to optimism about further experiments in that field without having identified the virus.

Following this, and considering the heavy pressure, it is not surprising that Loeffler was not in the contemplative mood required to publish sophisticated articles about the nature of the virus and to perform experiments to prove one or the other theory in this field. In addition, he was adherent to the bacteriological paradigm that had presented many successful discoveries up to that time and he did not see any convincing reason to reject this concept. Considering all the ramifications of his work, his research strategy, despite all the remaining questions and the epistemological uncertainty, becomes more comprehensible.

References

1. Abel R (1915) Friedrich Loeffler. Zentralbl Bakt 76: 241–245
2. Anonymus (1911) Die Tierseuchen-Kalamität. Berliner Abendpost, 27. April 1911
3. Beer J (1985) Friedrich Loeffler als Begründer der Virologie. Arch Exp Vet Med 39: 623–630

4. Beijerinck MW (1899) Ueber ein Contagium vivum fluidum als Ursache der Flekkenkrankheit der Tabaksblätter. Zentralbl Bakt Abt II 5: 27–33
5. Bernard C (1865) Introduction à l' étude de la médecine experimentale. J. B. Baillière, Paris
6. Bochalli R (1959) Friedrich Loeffler, Leben und Werk (Zur 75. Wiederkehr der Entdeckung des Diphtheriebazillus). Med Monatsschr 13: 59–62
7. Dittrich M (1963) Friedrich Loeffler (1852–1915) und die Virusforschung. Ein Beitrag zur Geschichte der Mikrobiologie. Gesch Naturw Techn Med (Beiheft NTM): 169–189
8. Dittrich NM (1975) Die Widerspiegelung der Robert-Koch-Schule in den Forschungsprogrammen der Greifswalder Medizinischen Fakultät. Ein Beitrag zur Geschichte der Mikrobiologie. Dissertation, Greifswald
9. Döhner L (1982) Friedrich Loeffler und die Entwicklung des Infektionsschutzes. Wiss Z EMA-Universität Greifswald (Gesellschafts-Sprachwiss Reihe) 31: 45–50
10. Döhner L (1983) Die Kochschen Postulate und die Virologie. Z Erbkrank Atm-Org 161: 21–24
11. Fraenkel-Conrat H (1988) Tobacco mosaic virus. In: Fenner F, Gibbs A (eds) Portraits of viruses. A history of virology. Karger, Basel, pp 124–146
12. Grafe A (1991) A history of experimental virology. Springer, Berlin Heidelberg New York Tokyo
13. Hatcher J (1978) Martinus Beijerinck. Gaz Inst Med Lab Sci. 22: 171–172
14. van Helvoort T (1991) What is a virus? The case of tobacco mosaic disease. Stud Hist Phil Sci 22: 557–588
15. Hüller H, Berndt HG, Dittrich M (1978) Das Wirken von Friedrich Loeffler in Greifswald. Arch Exp Vet Med 32: 313–318
16. Ivanovski DJ (1892) Ueber die Mosaikkrankheit der Tabakspflanze. Bull Acad Imp Sci St Petersburg 3: 67:70
17. Lechevalier H (1972) Dimitri Josifovich Ivanovski (1864–1920). Bacteriol Rev 36: 135–145
18. Loeffler F (1898) IV. Bericht der Commission zur Erforschung der Maul- und Klauenseuche bei dem Institut für Infektionskrankheiten in Berlin. Dtsch Med Wochenschr 24: 562–564
19. Loeffler F (1903) Bericht über die Untersuchungen der königlich Preussischen Commission zur Erforschung der Maul- und Klauenseuche in den Etatsjahren 1901 und 1902. Dtsch Med Wochenschr 29: 670–672, 685–687
20. Loeffler F (1903) Robert Koch. Zum 60. Geburtstag. Dtsch Med Wochenschr 29: 937–943
21. Loeffler F (1905) Die Schutzimpfung gegen die Maul- und Klauenseuche. Dtsch Med Wochenschr 31: 1 913–1 918
22. Loeffler F (1906) Ueber die Veränderung der Pathogenität and Virulenz pathogener Organismen durch künstliche Fortzüchtung in bestimmten Tierspezies und über die Verwendung solcher Organismen zu Schutzimpfungszwecken. Dtsch Med Wochenschr 32: 1 240–1 243
23. Loeffler F (1909) Die Serotherapie, die Seroprophylaxe und die Impfung bei Maul- und Klauenseuche und deren Wert für die Veterinärpolizei. Dtsch Med Wochenschr 35: 2 097–2 101
24. Loeffler F (1919) Filtrierbare Virusarten. In: Friedberger E, Pfeiffer R (eds) Lehrbuch der Mikrobiologie (mit besonderer Berücksichtigung der Seuchenlehre), vol 2. Gustav Fischer, Jena, pp 1 091–1 155
25. Loeffler F, Frosch P (1898) Berichte der 'Commission zur Erforschung der Maul- und

Klauenseuche bei dem Institut für Infectionskrankheiten in Berlin (I-III). Dtsch Med Wochenschr 24: 80–83, 97–100

26. Loeffler F, Uhlenhuth P (1901) Uber die Schutzimpfung gegen die Maul- und Klauenseuche, im Besonderen über die praktische Anwendung eines Schutzserums zur Bekämpfung der Seuche bei Schweinen und Schafen. Dtsch Med Wochenschr 27: 7–9

27. Loeffler F (1907) Neue Verfahren zur Schnellfärbung von Mikroorganismen, insbesondere der Blutparasiten, Spirochäten, Gonococcen und Diphtheriebacillen. Dtsch Med Wochenschr 33: 169–171

28. Lustig A, Levine AJ (1992) One hundred years of virology. J Virol 66: 4 629–4 631

29. Mochmann H, Köhler W (1984) Meilensteine der Bakteriologie. VEB Gustav Fischer, Jena

30. Mochmann H, Köhler W (1990) Friedrich Loeffler (1852–1915) – Wegbereiter der Bakteriologie und Virologie. Z ärztl Fortbild 84: 400

31. Moschell A (1994) Friedrich Loeffler (1852–1915). Ein Beitrag zur Geschichte der Bakteriologie und Virologie. Dissertation, Mainz

32. Mrowka F (1913) Das Virus der Hühnerpest ein Globulin. Zbl Bakteriol 67: 249–268

33. Röhrer H (1952) Zum 100. Geburtstag Friedrich Loefflers. Arch Exp Vet Med 6: 65–71

34. Schadewaldt H (1975) Die Entdeckung der Maul- und Klauenseuche. Dtsch Med Wochenschr 100: 2 355–2 395

35. Uhlenhuth P (1932) Das Lebenswerk und Charakterbild von Friedrich Loeffler. Zbl Bakteriol 125: I–XX

36. Uhlenhuth P (1952) Friedrich Loeffler als Forscher und Mensch. Gedenkworte und persönliche Erinnerungen anläßlich der 100. Wiederkehr seines Geburtstages (geb. 24.6.1852). Z Immunitätsforsch Exp Ther 109: 289–301

37. Wilkinson L (1974) The development of the virus concept as reflected in corpora of studies on individual pathogens. 1. Beginnings at the turn of the century. Med Hist 18: 211–221

38. Wilkinson L (1976) The development of the virus concept as reflected in corpora of studies on individual pathogens 3. Lessons of the plant viruses – tobacco mosaic virus. Med Hist 20: 111–134

Authors' address: Dr. H. P. Schmiedebach, Institut für Geschichte der Medizin, Walther-Rathenaustrasse 48, D-17487 Greifswald, Federal Republic of Germany.

The legacy of Friedrich Loeffler – the Institute on the Isle of Riems

W. Wittmann

Riems-Pharmaceuticals Ltd., Riems, Federal Republic of Germany

Summary. When starting the experiments on foot and mouth disease on the Isle of Riems in October 1910, Friedrich Loeffler could continue investigations that had been interrupted in 1907 by ministerial order. Loeffler's appointment to Berlin in 1913 and his sudden death in 1915 lead to the temporary cessation of work on the Riems. With high personal creativity and many years of seminal influence, Otto Waldmann carried Loeffler's ideas further, in the selection of themes and research strategy, making the Riems a world famous place of research. Some essential elements have determined life and research on the island for decades: the development of measures against epidemics, the conception of their application, the extension of research to new diseases of economic significance, the close contacts with the veterinary practice at all times, the presentation of results to experts and the stimulating discussions in the laboratory. I will try to briefly draw a bow covering the decades of different social conditions to the present and to suggest that Loeffler's ideas, which have been improved with the years, can affect many a current decision, even though differentiated individually.

*

The last hundred years of animal virus research often gave reason to remember its initiators and their achievements [7, 32, 60, 61]. Celebrations for different occasions mostly took place at Greifswald or on the Isle of Riems [6, 19, 42, 44]. In addition, Frankfurt an der Oder, Loeffler's native town, repeatedly honoured its great son, who was born in June, 24th, 146 years ago [3]. This contribution deals with a part of what we define as the legacy of Friedrich Loeffler, namely that part which has been linked for years with the name "Riems" for dealing with novel pathogens – the viruses – described for the first time in 1898 [28].

First one should start with Friedrich Loeffler: Some words Loeffler said about himself [24]: "... The remembrance of that time, when we were still working ... in the middle Robert Koch and we at his side, when we were faced almost daily with new miracles of bacteriology and when we – following our principal's shining example – were working from dawn to dusk hardly finding time to meet the physical requirements, the remembrance of that time I shall never forget. Surely

Fig. 1. Friedrich Loeffler (24.6.1852–10.4. 1915) (Archives of Riems)

we learned in those days what it means to observe and to work exactly and to follow a fixed target with energy." (Fig. 1)

Uhlenhuth, Loeffler's most recent co-worker said in memoriam [60]: "... when he succeeded in finding a new staining method ... he could be glad like a child to demonstrate us his specimen preparations under the microscope putting back his glasses and exclaiming enthusiastically in his vivacious nature: 'Candy, candy, Gentlemen!' "

A little book about the University of Greifswald mentions him as "... a lively personality with keen very friendly glittering eyes hidden behind glasses, the lower lip pushed forward somewhat gruffly out of the blond trimmed beard, the manner of speaking short, in Berlin dialect. Wherever he was he created around himself a ring of security, clarity and gladness ..." [30]

Loeffler's person was also the subject of some doctoral theses. One of the latest dissertations [33] reports: "... so he surely succeeded in his most important discoveries mainly due to his attitude towards science, his creativity his obsession with science and his courage to take a risk. In spite of brilliant achievements being the results of strenuous work, he had to overcome first of all the considerable opposition of the competent colleagues and the small-mindedness of the authorities particularly in his research dealing with foot and mouth disease..." We can imagine a little how Friedrich Loeffler arranged his working day, how he formed new ideas and organised the experimental basis for them mentally. We can feel him being glad about good results, how he made every efforts to share this gladness with his collaborators and how he had a longing for discussing with them, before writing down the results for the next publication. Without wanting to place Loeffler into the present I think that many of his activities not only impressed friends and colleagues but also many a researcher of following generations who could only read what the Privy Councillor was doing. – This statement may be a little idealised, as young scientists often are no longer interested in things being older than a week, as Sidney Brenner once said [8]. When working for the preparation of the Riems 50th anniversary in 1960, I became very much aware of working in the world's first virological institute. Sometimes, however, you remember ancestors of your subject more vividly and this has not changed until now, as Walter Plowright underlined last year [37].

In the following, one should emphasise one part of the legacy, referring to selected aspects of the Riems research, the more so as some of these results substantially influenced the development of the institute. Again a statement by Loeffler is placed at the beginning [27]: "...Robert Koch's example showed us how to take effective measures for the control of agents based on the knowledge of them and their biology." After October 10, 1910, the day when the experiments of foot-and mouth-disease (FMD) was restarted with the lymph which arrived at the Riems from Vickovo, he realised this principle vehemently. The classification of FMD as a disease in which protective antibodies are produced led to the development of an immune serum, its first administration – alone and together with FMD virus, named "seraphtin" – showed that the complex of prophylactic, therapy and control of epidemics is inseparable [25].

Astonishingly these developments were fiercely attacked from veterinary circles and Loeffler's results were frankly regarded as irrelevant [50]. Loeffler got used to such objections and was trained in polemics [23].

He felt angry repeatedly that not sufficient immune serum was available, and that it was so expensive. Loeffler himself said [26]: "... At present the fairly high price still interferes with the common administration...but I hope that it will become cheaper in the course of time". Loeffler's hope came true but he did not live to see it. "When I was appointed director of the Robert Koch Institute for Infectious Diseases, I ceased working on the Isle of Riems..." [27]. Loeffler died in 1915 and was buried in Greifswald.

The experiments on the Riems also ceased. The demise of the whole institute was nearby, had it not been for Dr. Nevermann from the Berlin ministry who made – as in Loeffler's lifetime – a strong effort to promote and protect Loeffler's idea of FMD research on the Isle. In May 1919, Dr. Otto Waldmann, a veterinarian employed at the Berlin Veterinary Faculty, was appointed assistant to the district veterinarian on the Isle of Riems. Waldmann filled himself and his rapidly growing staff of co-workers with enthusiasm. As early as 1926, Kurt Wagener, one of the coworkers on the Isle of Riems, appraised this activity as follows [64]: "...The scientific progress culminated economically in an extraordinary drop in the costs for serum production and thus in a strong reduction of the serum price from more than 150 to about 50 Marks per litre...The few hundred litres of serum which had formerly been produced per year can now be produced within days...Today the research institute has two laboratory buildings with modern equipment, where the scientific work for the further research of foot and mouth disease is carried

out. The stables were enlarged so that at present 300 adult cattle can be kept..."
Scientific progress during these years was characterised by the transmission of
the virus to the guinea pig [68], evidence of FMD virus' diversity [67, 59], the
first attempts to grow the virus in tissue culture [15], and the serotype diagnosis
using the complement-fixation reaction [58]. Further work during those years
was dedicated to basic investigations about the FMD virus in order to obtain an
inactivated vaccine as soon as possible and to improve measures of epidemic
control for inhibiting the spread of the disease [12, 40, 72, 69].

In the 1930s, the Riems efforts showed first successes when in different attempts a combination was found. This included own experience, French results
about the possibility to inactivate FMD virus with formaldehyde [62], and the finding of a Danish veterinarian [51] about adsorbing the virus to aluminium hydrox-
ide. Due to these efforts, the first efficient inactivated FMD-vaccine was produced
[70]. Very soon after having tested it under experimental conditions, an excellently
organised and evaluated field trial [66] was carried out in 15,200 cattle, 1,600
sheep and 320 goats. The following results [29, 13] were reported: "...the vaccination did not show any disadvantages, neither for the single animal nor for the whole
population, that the protection was fully developed 6 to 12 days after vaccination
and that vaccinated animals were protected longer than 3 months, even in cases of
massive contact infection... In the control of foot and mouth disease the Riems vaccine will take away the fright from this devastating disease, as soon as it can be used
comprehensively."

Also the press reported repeatedly and informed that: "... 2.5 million
Marks will be made available in order to extend the Riems institute" [53]. These
means were used to finish the second extensive building phase of Waldmann's
Riems and in 1940, the main building was completed. In those days, it was the
domicile of the microbiological division (Prof. Traub), the divisions of pathology (Prof. Röhrer), chemistry (Prof. Pyl) and production (Prof. Möhlmann)
(Fig. 2).

After several discussions, the *Office International des Epizooties* in Paris
recommended the prophylactic use of the Riems FMD vaccine and the teams on
the Riems were busy with the continuous improvement of their vaccine [35, 71].
As early as 1942 the annual production was 80,000 litres of mono- or 50,000
litres of bivalent vaccine, being sufficient for 1.5 million or 900,000 cattle, respectively [65]. However the maintenance of this progress become complicated
during World War II, especially in 1943 to 1945 [11, 74]. The whole installation
was disassembled after the war and on one of the laboratory walls an unknown
wrote "Research is finished". Thus, many of those having lived and worked on
the Riems for years with high creativity and propensity for research left the island
[44, 31]. Together with some co-workers Waldmann went to Argentina but returned in 1953. Friends and colleagues spoke at his grave in 1955. W. Nußhag, the
long-time neighbour of the Waldmann family in Greifswald, one of my teachers
at the Berlin University, said at the grave [34]: "... this man was able to build the
first, the greatest and the finest institute for virus research, the example for all
others ..." (Fig. 3)

Fig. 2. The main building of the institute (erected 1940) on the Isle of Riems; in front of the institute the sculpture "The cow" (created in 1960) by F. Cremer, Berlin (Archives of Riems)

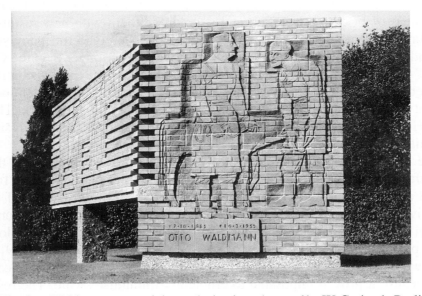

Fig. 3. The Otto Waldmann memorial near the institute (created by W. Grzimek, Berlin, 1960)
(Archives of Riems)

Fig. 4. The rector of the Greifswald University (Prof. Werli) presenting to Prof. Röhrer the letter of appointment to the post of Professor for Virology at the medical faculty on October 10, 1960 (Archives of Riems)

The poorly equipped laboratories became the "Institute for the Control of Foot and Mouth Disease". An increase in production and the start of reconstruction began because of an order of the soviet military administration, beginning in November 1948. During these years the efforts and devoted work of each one wishing to preserve the laboratory and to maintain the FMD control according to the ideas of Friedrich Loeffler, and to improve them with new results, went to the limits of endurance. The motive could have been read for many years on the window in the middle of the foyer – "Our will was stronger than the German misery" – was a daily hint for each co-worker as to his own expected contribution to the rebuilding of the laboratory after 1945. It called each guest's attention to the fact that everything now in existence was achieved only by hard work. This was led by Heinz Röhrer who returned to the Riems in 1948 and acted as President of the institute until 1970 [48, 21] (Fig. 4).

The development of the Riems FMD concentrated vaccine during the first years after World War II enabled a reduction of the immunising dose from 30 to 5 ml per cattle and the increase in vaccine production on the Isle of Riems, respectively, represented the basis of the annual prophylactic vaccination in the German Democratic Republic (GDR) [43]. "From 1961 to 1965 a pilot plant for the production of FMD-tissue culture vaccines was established" [49]. Great efforts were put into developing a FMD live attenuated vaccine for application in pigs, because it proved impossibe to efficiently immunise swine. The results of these experiments were unsatisfactory [16, 52].

In the early 1950s, due to the political partition of Germany, it became increasingly obvious that also in the Federal Republic an FMD vaccine production had to be secured since some epizootics had already caused heavy damage. Well-known manufacturers carried out this vaccine production. Nevertheless there was also a need to conduct research on FMD and other virus infections of animals, and an appropriate institute was founded in Tübingen, with Erich Traub as the first president. Because he was a former Riems co-worker, it stood to reason that he would bring considerable experience and Waldmann's ideas to the construction and profile of the Tübingen Research Institute.

Cattle were vaccinated in many European countries. This was an essential contribution to the containment of FMD in Europe and fundamental to the fact that since July in 1988, East and West Germany are officially free from FMD [20], an aim achieved 50 years after the first field trial with Riems' FMD-adsorbate-vaccine in 1938. It might also well be to Loeffler's liking if we recall his words of 1914 [27]: "...since I have been engaged in this research since 1896, I am keenly interested in it and I wish with all my heart that it will be always promoted, for the benefit and the welfare of German agriculture..."

In early 1992, vaccination against FMD was stopped within the European Community. Apart from the changes that had already taken place at the Riems Institute, it had consequences [73], which gave reason to speak about a possible end of the Loeffler tradition [5].

The aim of research at the Riems, however, was not just vaccination in general but the elimination of the threats caused by FMD using a vaccine, always assisted by veterinary sanitary measures until the disease was eradicated and the agent eliminated. This fundamental principle was also in the Loeffler tradition; with regard to hog cholera it was defined by Röhrer, who said [45]: "...if once the disease has been eradicated completely and there is no or little danger of introducing it into a sanitised area then vaccination could be consequently dispensed with ...and the measures of control could be again solely veterinary sanitary ones."

However, Loeffler's legacy is more than FMD research alone. As early as 1912, he postulated [26]: "For an extremely great number of diseases of man and animals, such as yellow fever, rabies, hog cholera, fowl plague, equine plague, pox diseases, to count only a few no causative agent had been found. As research showed they all are caused by a virus..." And so in the late twenties investigations on hog cholera started on the isle. Later, almost exclusively for economical reasons, the range of research was extended to virus diseases which required an etiological and diagnostical cleaning up and a control strategy. By the words of W. Nußhag [34] Otto Waldmann was able to continue the legacy of Loeffler over all the years. In 1955 he said: "...always he underlined the practicable application of the new findings both a sophisticated method of virus research and the construction of straw-huts to prevent influenza of pigs..."

Thus in the 1930s the Riems institute very quickly became not only a consultancy office for cattle holders but also for the pig breeders, for the horse breeders and owners, as well as a source for poultry farmers in the case of new diseases (Table 1).

The extension of the Riems' research tasks continued after the work was resumed in 1948. In Riemserort a production plant was built, particularly for

Table 1. Scientific items of the Institute on the Isle of Riems until 1948

Viral infections (Beginning of elaboration)	Items	Co-worker
Hog cholera (1930)	diagnosis, latent infection	David, Schwarz
	pathology, histology	Röhrer, Waldmann
	cytology of blood	Nagel
Influenza of pigs (1932)	etiology	Köbe, Waldmann
	diagnosis	Schmidt
	epidemiology, eradication	Vogt, Radtke, Hein
Cough of horses (1934)	etiology	Waldmann, Köbe
Bronchitis of cattle (1935)	etiology	Waldmann, Köbe
Pneumonia of calves (1937)	etiology	Nagel
Infect. anemia of horses (1938)	experimental transmission	Köbe
Fowl plague (1943)	pathogenesis, diagnosis	Dinter, Röhrer
	vaccine	Traub
Mouse poliomyelitis (1944)	histopathology	Röhrer

Fig. 5. Building of the CVV production (1956) in the village Riemserort, where some laboratories of the production division of the "Friedrich-Loeffler-Institut" had been situated; since 1991 this building is a part of the "Riems Pharmaceuticals Ltd" (Archives of Riems)

crystal violet vaccine (CVV) against hog cholera; it started production in 1956 ([10, 44, 56]; Fig. 5). Equine infectious anaemia and fowl plague remained major topics in the Riems research programme [39, 17]. New research, traditionally initiated from practice, in order to assist the veterinary practitioner and the diagnostician as well as to improve basic knowledge, was initiated (Table 2). At the age of 85 years, Röhrer summed this philosophy up as follows [47]: "...The foot and mouth disease vaccine as well as the crystal-violet-vaccine against hog cholera are striking examples for a successfully completed systematic research carried out with a strict strategy. They result from a lively relation to the veterinary practice, which has always been exercised by the institute. That is also true for ...other scientific achievements of the Riems institute. They have almost all been elaborated in close interweaving of basic and applied research and technology as well as in their reciprocal fertilisation. In this sense the Riems institute has been working since its foundation flexible in its inner structure and its interdisciplinary co-operation..."

Röhrer's successors struggled to maintain the Riems tradition. More than ever, agriculture and the authorities demanded not only research results about the current epidemics within the highly industrialised animal production in the GDR but

Table 2. Scientific topics and vaccine production in the "Friedrich-Loeffler-Institute" on the Isle of Riems, 1948–1970[a]

Main scientific items	Vaccine production
Aujeszky's disease of pigs	FMD-vaccine
Borna disease	based on
Pustular dermatitis	*aphthes from infected cattle
Enzootic bovine leukosis	*tissue culture
Foot-and mouth-disease	
Fowl plague	Hog cholera
Hog cholera	*crystal-violet-vaccine
Inf. bovine rhinotracheitis	
Influenza of pigs	
Inf. laryngotracheitis of poultry	
Inf. bronchitis of poultry	
Infections of laboratory animals	experimental vaccines
Mucosal disease	FMD (live vaccines)
Ornithosis	*egg-adapted virus
Parainfluenza III inf. of calves	*mouse adapted virus (neurotropic)
Teschen disease	*tissue culture adapted virus
Rabies	
Stomatitis papulosa	
Talfan disease	
Transmissible gastroenteritis of pigs	

Unlike in Table 1, it is not possible to name the coworkers in Tables 2 and 3; most of the coworkers are listed in the chapter references or in the papers of some of the cited authors of the Institute on the Isle of Riems

Table 3. Scientific topics and vaccine production in the "Friedrich-Loeffler-Institute" on the Isle of Riems, 1971–1990[a]

Main scientific items	Vaccine/diagnostic production
Application of vaccines	vaccine production
*by aerosol (CSF, Aujezsky ...Erysipelas of swine)	*FMD (BHK21-tissue culture vaccine)
*oral delivery systems (TGE, SVC)	*Hog cholera strain C in rabbits
Enzootic bovine leukosis	
Foot-and-mouth-disease	strain C in tissue cultures
Hog cholera (CSF)	*Marek's disease
Inf. bovine rhinotracheitis	*Transmissible gastroenteritis
Marek's disease	*Infect. bovine rhinotracheitis
Modern diagnostical systems (ELISA)	*Parainfluenza III
Mucosal disease	*Rabbit haemorrhagic disease
Parainfluenza III inf. of calves	diagnostics
Transmissible gastroenteritis of pigs	*several immunofluorescence sera
*Corona and corona-like infections	*BLV test kit (AGPT)
Viral diseases of fish (e.g. spring viraemia of carp, SVC)	
Swine vesicular disease (SVD)	vaccines in experimentally forms
	*inactivated SVD vaccine
	*BEI-inactivated FMD-vaccine
*Oil-adjuvanted FMD-vaccine for pigs	*SVC oral vaccine

[a]Some items were studied in close co-operation with the "Institute of Vaccines" Dessau, Central Institute of Cancer Research and Central Institute of Molecular Biology of the Academy of Sciences in Berlin-Buch; Faculty of Pharmacology of the Martin-Luther University in Halle etc. and several Veterinary Research Institutes in Bulgaria, Czechoslovakia, Hungary, Poland, the (former) Soviet-Union and Romania

also the production of vaccines and diagnostics on the island and new technology in application of vaccines [63]. It was possible to present to the Academy of agricultural sciences of the GDR good results in most of the cases despite material and technical problems (Table 3). Results from the Riems Institute were highly regarded. However, for safeguarding of the secrets of production and the most of the epidemiological details about viral diseases in animal production, many of the results remained unpublished or were only allowed to be published in German and only rarely in foreign specialist periodicals [2, 4]. Because production was beneficial for the "economy" of the institute warning words like "today the Isle of Riems is practically a people-owned vaccine plant" did obviously not influence the existing plans [46]. Soon, however, we realised that due to this trend and other restrictions we were no longer recognised world-wide and even failed to answer comprehensively questions about the products of the Isle, even practical questions from veterinarians in the country. Emphasis on the exclusively applied research resulted in neglect of urgently needed basic work [57]. Even when a molecular biology team was established application had priority [18]. The legacy of Friedrich Loeffler and the worldwide reputation based in the era of Waldmann and Röhrer faded away. In 1985, Zvonimir Dinter, who had been working on the Riems from 1943 to 1945, wrote [9]: "In 1960, the Riems research institute cele-

Fig. 6. Dr. Zvonimir Dinter participating in the scientific symposium "50 years of Riems" (midst, with pipe). On the left: Dr. Hansen (Stockholm) and the Russian interpreter. On the right: Dr. Moosbrugger (Basel). In the foreground from the left to right: Dr. Szent-Ivanyi (Budapest), Dr. Bakos (Stockholm), Prof. Dr. Rubarth (Stockholm) and Dr. Tomescu (Bukarest). In the background: Collaborators of the Riems Institute and Dr. Lieschke (Academy of Agricultural Sciences, Berlin; Archives of Riems)

brated its 50th anniversary...and many virologists from here and abroad (including myself) accepted the invitation... It was a successful meeting in every respect and in this, I saw Röhrers attempt to foster the international association of scientists and to establish contacts. Shortly thereafter the curtain of isolation went down over the Riems and this became a permanent condition. I do not know why this was necessary but I said to myself: Riems *mon amour* – as if I had lost a dear friend." (Fig. 6)

The political events of 1989 provided the opportunity to return Riems to what it was. In this complicated situation, the former president himself stood frequently by the management on the island. Future trends were suggested by the Scientific Advisory Council [73]: "Not only because of the vaccine production which has to be removed to the private business but also because of the far-reaching restraint of formerly significant virus diseases research at the FLI needs a new orientation...The Council recommends to bring together both facilities working in the field of virus research on the Isle of Riems sooner or later..."

In 1991 the production units was separated and privatised. The insular part became the "Federal Research Centre for Virus Diseases of Animals" (the first president was Prof. Dr. Volker Moennig; since 1994 Prof. Dr. Thomas C. Mettenleiter) and is now to become the only location of veterinary research in this field in Germany. It is self-evident that even today one is faced with Loeffler's ideas of research on the Riems, particularly when reading: "...The main task of the institute is the development of diagnostics and vaccines..." and coincidentally you are quite close to Loeffler's striving to know the causative agents ("...Robert Koch showed us clearly how to can take effective measures for the control of them based on the knowledge of them and their biology..."), even if using the methods 100 years after Loeffler, according to Mettenleiter who emphasized: "... to define the causative agents referring to their molecular biology and to trace out the gene factors being responsible for the disease causing properties of the viruses..." [36].

Finally, one must mention the sites of memory and tribute to the founder of animal virus research on the Isle of Riems. There are memorial stones, busts and pictures. These honour those Riems veterinarians who, together with Friedrich Loeffler, the Doctor of veterinary medicine *honoris causa* of the Gießen Veterinary Faculty, take prominent place in the annals of veterinary medicine. It is possible to show the subjects dealt with. And so the "guinea pig monument" (Fig. 7) and the "cow" have to be placed into the set of artistic monuments, too. They are not only presentations of animals in general but illustrate that only by using these creatures was it possible for work on research on the health of domestic animals. This idea continues in the sculptures, designed by W. Grzimek, which confront life and death artistically and call on the researcher to protect life. This chain of thoughts leads to one of the paintings of H. Neubert, showing the origin of modern virus propagation in cell culture and integrating the thought of minimising tests in animals by using such systems. The cell collection on the Isle of Riems,

Fig. 7. The sculpture "guinea pigs" created by F. Cremer, Berlin (Archives of Riems)

established in 1973 [41], "...is of high significance for virus research and unique in the whole Federal territory..." as the scientific council stated in 1991 [73]. The series of paintings made by H. Neubert has held an eminent place for years. It shows the level that virus research had reached 50 years after the beginning of Loeffler's activities on the Riems [38]. It also illustrates the progress of the recent past achieved by the creative work of researchers all over the world, nowadays the way to characterize the structure and function of each molecule in order to increase basic knowledge as well as to improve diagnostics and vaccines. The painting showing the electron microscope shall be mentioned, for it was on the Riems where the first electron microscope constructed by Ruska was installed and by means of which Ardenne and Pyl [1] tried to visualise FMD virus for the first time (Fig. 8). Therefore, this painting is more than simply a snapshot. It highlights that scientists of the Riems contributed to the characterisation of Loeffler's FMD virus, using electron microscopy [54], characterisation of the nucleic acids [17], and sedimentation analysis of the FMD virus [22] in the 1960s (Fig. 9).

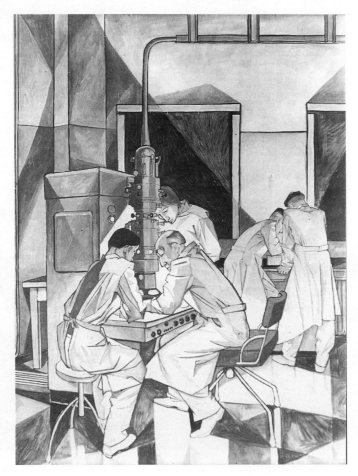

Fig. 8. The picture "electron microscope", painted by H. Neubert, in the foyer of the main building of the Institute on the Isle of Riems (Archives of Riems)

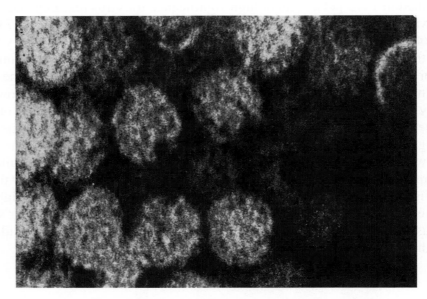

Fig. 9. Electron microscopy of FMD virus (original 640 000 fold) negative staining (Schulze and Gralheer – see also [54]; Archives of Riems)

Fig. 10. The houses on the Isle of Riems, in which F. Loeffler and his coworkers were working during the first years (Archives of Riems)

Now we have returned to that virus whose first description 100 years ago by Loeffler and Frosch brought us together and who chose the Isle of Riems as a research station (Fig. 10).

I close with the words from a contemporary of Loeffler, the chemist Wilhelm Ostwalds who said [55]: "And the progress of science coincides with a steadily growing number of reliable signposts". Some of the numerous Riems signposts have been presented, others not. Researchers at the institute on the Isle of Riems have ample opportunities to find and to place new signposts.

Acknowledgements

The author should like to thank the President of the Federal Research Institute on the Isle of Riems, Prof. Thomas C. Mettenleiter for the opportunity to use library and archive; thanks go also to Mrs. B. Riebe, librarian, and Mr. H. Stephan, photographer, from the same institute, for helpful assistance.

References

1. Ardenne M, Pyl G (1940) Versuche zur Abbildung des Maul- und Klauenseuchevirus mit dem Universal-Elektronenenmikrokop. Naturwissenschaften 28: 531–532
2. Anonymous (1971) Anordnung zum Schutz der Dienstgeheimnisse vom 6. Dez. 1971. Privatbesitz
3. Anonymous (1984) Gemeinsame Tagung von Medizinern und Historikern anläßlich des 100. Jahrestages der Entdeckung des Diphterieerregers. Neuer Tag 23. 3. 1984 (Wochenendbeilage S 6)
4. Anonymous (1986) Ordnung über die Öffentlichkeitsarbeit im VEB Kombinat Veterinärimpfstoffe vom 1. Jan. 1986. Privatbesitz
5. Bartels TH (1992) Aus für Riemser MKS-Forschung. Ostseezeitung 3. Jan. 1992
6. Beer J (1985) Friedrich Loeffler als Begründer der Virologie. Arch Exp Veterinärmed 39: 623–630
7. Bocchalli R (1959) Friedrich Loeffler, Leben und Werk (Zur 75. Wiederkehr der Entdeckung des Diphteriebacillus). Med Mschr 13: 59–62
8. Brenner S cited by Plowright [37]
9. Dinter Z (1985) Auf Riems. Tierärztl Umschau 40: 1 006–1 009
10. Fuchs F (1959) Die Entwicklung der Immunität nach der Impfung mit Kristallviolettvakzine gegen Schweinepest. Mber Dtsch Akad Wiss 1: 650–659
11. Geissler E (1998) Wurde Himmler hinters Licht geführt? In: Geissler E (Hrsg) Biologische Waffen nicht in Hitlers Arsenalen. LIT-Verlag, Münster, pp 231–246
12. Gomolka G (1931) Beitrag zur Antigengewinnung für die aktive Immunisierung gegen Maul- und Klauenseuche mittels abgetötem Virus. Dissertation, Berlin
13. Haan W (1938) Nachwort zu vorstehendem Artikel des Veterinärrat Dr. Mass. Berl Münch Tierärztl Wochenschr 1938: 509–510
14. Hantschel H (1961) Ein infektiöses Prinzip mit Ribonukleinsäure-Charakter aus Mäusegehirnen, die mit neurotrop modifiziertem MKS-Virus infiziert waren. Arch Exp Veterinärmed 17: 263–265
15. Hecke F (1931) Die Züchtung des Maul- und Klauenseucheerregers. Dtsch Tierärztl Wochenschr 39: 258–261
16. Heinig A (1964) Tot- und Lebendvirusimpfstoffe zur Bekämpfung der Maul- und Klauenseuche (MKS). Sitzungsber Akad Landwirtsch XIII Heft 4

17. Heinig A, Schmidt U (1954) Untersuchungen über das Herfordshire-Virus. Arch Exp Veterinärmed 8: 517–521
18. Heinrich HW (1987) Der Einsatz gentechnischer Methoden in der Virologie. Arch Exp Veterinärmed 41: 660–664
19. Hüller H, Berndt HG, Dittrich M (1978) Das Wirken von Friedrich Loeffler in Greifswald. Arch Exp Veterinärmed 32: 313–318
20. Kramer M, Jentsch D, Kaden V, Schlüter H, Zwingmann W (1997) Die Maul- und Klauenseuche-Situation in Europa und der Welt. Amtstierärztl Dienst Lebensmittelkontr 4: 196–202
21. Liebermann H, Kaden V, Beyer J (1992) Professor Heinz Röhrer zum Gedenken. Tierärztl Umschau 47: 654–655
22. Liebermann HT (1968) Physikalische Charakterisierung des neurotrop modifizierten Maul- und Klauenseuche-Virus. Dissertation, Greifswald
23. Loeffler F (1902) Bericht der Königlich-Preußischen Commission zur Erforschung der Maul- und Klauenseuche (Geheimrat Loeffler und Stabsarzt Uhlenhuth) über das Bacellische Heilverfahren. Dtsch med Wochenschr 1902: 245–249
24. Loeffler F (1903) Robert Koch zum 60. Geburtstage. Dtsch med Wochenschr 29: 938–943
25. Loeffler F (1910) Die Serotherapie, Seroprophylaxe und die Impfung bei Maul- und Klauenseuche und deren Wert für die Veterinärpolizei. Berl Tierärztl Wochenschr 1910: 870–873
26. Loeffler F (1912) Über den heutigen Stand der Erforschung der Maul- und Klauenseuche. Ref. Sitzung Dtsch Landwirtsch Ges 1912
27. Loeffler F (1914) Die Verbreitung der Maul- und Klauenseuche und der gegenwärtige Stand ihrer Bekämpfung. Arch wiss prakt Tierhk 40: 308–323
28. Loeffler F, Frosch P (1898) Berichte der Kommission zur Erforschung der Maul- und Klauenseuche bei dem Institut für Infektionskrankheiten in Berlin. Centralblatt Bakteriol Abt I Orig 23: 371–391
29. Maas A (1938) Die aktive Immunisierung mit Riemser Maul- und Klauenseuchevakzine nach Waldmann und Köbe. Berl Münch Tierärztl Wochenschr 1938: 477–508
30. Malade T (1938) Aus einer Kleinen Universität. Greifswald, München, Berlin 1938
31. Mayen F (1995) Die Geschichte der Tierseuchenbekämpfung beim Rind in Argentinien mit besonderer Berücksichtigung der MKS und der Rolle Otto Waldmanns in deren Bekämpfung. Dissertation, Med Vet Freie Universität Berlin
32. Mochmann K, Köhler W (1980) Friedrich Loeffler – 1852–1915 – Wegbereiter der Bakteriologie und Virologie. Z ärztl Fortbild 84: 400–406
33. Moschell A (1994) Friedrich Loeffler (1852–1915) – Ein Beitrag zur Geschichte der Bakteriologie und Virologie. Dissertation, Joh Gutenberg Univ Mainz
34. Nußhag W (1955) Worte des Gedenkens für Otto Waldmann. Manuskript in Privatbesitz aus dem Nachlaß der Tochter, Dr. E. Nagel.
35. OIE (1939) Resolution des Internationalen Tierseuchenamtes "Riemser Vakzine gegen Maul- und Klauenseuche". Sitzung vom 30./31.5. 1939 in Paris
36. Pechmann EV (1998) Grösser als die Winzigsten. J Ernst-Moritz-Arndt Univ 9: 4–5
37. Plowright W (1997) Three revolutions in 50 years of microbiology: a personal view of their effects and lessons (Norbrook lecture). Vet Rec 22: 296–301
38. Pommeranz-Liedtke G (1960) Zehn Wandbilder von Hans Neubert spiegeln Sinn und Praxis der Virologie. Arch Exp Veterinärmed 14: 841–855
39. Potel K (1952) Zur Histopathologie der Infektiösen Anämie der Pferde. Arch Exp Veterinärmed 6: 118–131, 132–156, 363–374

40. Pyl G (1931) Adsorptionsversuche mit Maul- und Klauenseuchevirus in Pufferlösungen. Zbl Bakteriol Abt I Orig 121: 10–19

41. Riebe R (1978) Aufgaben und Zielstellung des RGW-Referenzzentrums für permanente Zellinien am Friedrich-Loeffler-Institut für Tierseuchenforschung Insel Riems. Arch Exp Veterinärmed 32: 449–453

42. Röhrer H (1952) Zum 100. Geburtstag Friedrich Loefflers. Arch Exp Veterinärmed 6: 65–71

43. Röhrer H (1957) Die Riemser Maul- und Klauenseuche-Konzentratvakzine. Dtsch Tierärztl Wochenschr 64: 69–72

44. Röhrer H (1960) 50 Jahre Insel Riems. Arch Exp Veterinärmed 14: 713–763

45. Röhrer H (1965?) Die Bekämpfung der Schweinepest in der Deutschen Demokratischen Republik. Unveröffentl. Vortrag, Privatbesitz aus dem Nachlaß H. Röhrers

46. Röhrer H (1985) Erleben, erfahren, erkennen – Akademiemitglied Heinz Röhrer gibt zu Protokoll. Spectrum 16: 7–9

47. Röhrer H (1990) Aus der Ansprache auf dem Ehrenkolloquium zum 85. Geburtstag von H. Röhrer am 23. März 1990 auf der Insel Riems. Vollständiges Manuskript, Privatbesitz

48. Röhrer H (1992) Rede von Herrn Professor Röhrer anläßlich des Ehrenkolloquiums zu seinem 85. Geburtstag am 23. März 1990. Tierärztl Umschau 47: 926–929

49. Röhrer H, Liebermann H (1969) Die Entwicklung des Riemser Forschungsinstitutes in der Deutschen Demokratischen Republik. Mh Vet Med 24: 694–701

50. Schmaltz R (1911) Die Tierseuchen-Kalamität, Wertlosigkeit der Schutzimpfungen. Berl. Abendpost 98, 27.4.1911

51. Schmidt S (1936) Immunisierung von Meerschweinchen gegen drei verschiedene Typen von Maul- und Klauenseuchevirus vermittels eines trivalenten Aluminiumhydroxydadsorbates. Z Immun Forsch 88: 91–98

52. Schmidt U (1960) Über Versuche zur Vakzinierung von Rindern mit eiadaptiertem lebendem Maul- und Klauenseuchevirus. Arch Exp Veterinärmed 14: 568–576

53. Schünemann K (1938) Serum gegen Maul- und Klauenseuche. Wochenblatt 2.7. 1938

54. Schulze P, Gralheer H (1964) Untersuchungen zur Feinstruktur des Maul- und Klauenseuchevirus. Arch Exp Veterinärmed 18: 1 449–1 458

55. Strich M, Hossfeld P (1987) Wissenschaft im Zitat. Bibliogr Institut Leipzig S 23

56. Tesmer S, Werner P, Glaner M, Lange B, Gentner F (1985) Impfstoffproduktion am Friedrich-Loeffler-Institut Insel Riems. Arch Exp Veterinärmed 39: 684–691

57. Thalmann G, Wittmann W (ed) (1990) Friedrich-Loeffler-Institut für Tierseuchenforschung Insel Riems – 80 Jahre im Dienst der Tiergesundheit. Ostseezeitung, Verlag & Druck GmbH Greifswald

58. Traub E, Möhlmann H (1943) Typbestimmung bei Maul- und Klauenseuche mit Hilfe der Komplementbindungsprobe. Zbl Bakt Abt I Orig 150: 298–299 u. 300–324

59. Trautwein K (1927) Die Pluralität des Maul- und Klauenseuchevirus – Habilschrift – Arch wiss prakt Tierheilk 56: 505–535

60. Uhlenhuth P (1932) Das Lebenswerk und Charakterbild von Friedrich Loeffler – Gedenkrede zu seinem 80. Geburtstag. Zbl Bakt Abt I Orig 125: I–XX

61. Uhlenhuth P (1952) Friedrich Loeffler als Forscher und Mensch. Z Immun Forsch Exp Ther 109: 289–301

62. Vallee H, Carre H, Rinjard P (1925) Sur l' immunisation antiaphtheuse par le virus formole. Rev Gen Med Vet 35: 129–137

63. Vogel S (1999) Die Insel Riems als Wissenschaftsstandort – Vergangenheit, Gegenwart, Zukunft – Forschung auf dem Riems von 1970 bis 1991. Dissertation, Med Vet (in preparation)

64. Wagener K (1926) Die Insel Riems. Heimatkalender für die Kreise Greifswald und Grimmen 11: 21–25
65. Waldmann O (1942) Über die zukünftige Gestaltung der Bekämpfung der Maul- und Klauenseuche mit Hilfe der Vakzinierung. Berl Münch Tierärztl Wochenschr 1942: 221–227
66. Waldmann O, Köbe K (1938) Die aktive Immunisierung des Rindes gegen Maul- und Klauenseuche. Dtsch Tierärztebl 5: 318–320
67. Waldmann O, Mayr K (1924) Experimentelle Untersuchungen über die Richtigkeit der französischen Auffassung von der Pluralität des Maul- und Klauenseuchevirus. Berl Tierärztl Wochenschr 40: 37–38
68. Waldmann O, Pape J (1920) Die künstliche Übertragung des Maul- und Klauenseuchevirus auf das Meerschweinchen. Berl Tierärztl Wochenschr 1920: 519–520
69. Waldmann O, Reppin K (1935) Experimentelle Untersuchungen zur aktiven Immunisierung gegen Maul- und Klauenseuche. Z Inf Krkh Haustiere 47: 283–322
70. Waldmann O, Köbe K, Pyl G (1937) Die aktive Immunisierung der Rinder gegen Maul- und Klauenseuche mittels Formolimpfstoff. Zbl Bakteriol Abt I Orig 138: 461–468
71. Waldmann O, Pyl G, Hobohm KO, Möhlmann H (1941) Die Entwicklung des Riemser Adsorbatimpfstoffes gegen Maul- und Klauenseuche und seine Herstellung. Zbl Bakteriol Abt I Orig 148: 4–19
72. Waldmann O, Trautwein K, Pyl G (1931) Die Persistenz des Maul- und Klauenseuchevirus im Körper durchseuchter Tiere und seine Ausscheidung. Zbl Bakt Abt I Orig 121: 19–32
73. Wissenschaftsrat (1991) Stellungnahme zu den außeruniversitären Forschungseinrichtungen auf dem Gebiet der ehemaligen DDR im Bereich Agrarwissenschaften Teil III.2: 86–97
74. Wittmann W, Gadebusch-Bondio MC (1998) Die Reichsforschungsanstalt Insel Riems – auch Ort Kriegsgefangener des Dritten Reiches. 5. Tagung der FG Geschichte der Veterinärmedizin der DVG, Nov. 1997, Hannover, pp 181–192

Authors' address: Dr. W. Wittmann, Riems-Pharmaceuticals Ltd., An der Wiek 7, D-17498 Insel Riems, Federal Republic of Germany.

The post-Loeffler-Frosch era: contribution of German virologists

R. Rott

Institut für Virologie, Justus-Liebig-Universität Giessen,
Giessen, Federal Republic of Germany

Summary. This presentation dealt with the contributions of German virologists in the rapid development of virology following the Loeffler-Frosch era. Thereby, only research was included which was undertaken within German institutions, even though guest scientists from other countries or international cooperative efforts have in some cases contributed to the work. Contributions to the field of veterinary virology were not considered here, since this topic was treated separately during this centennial symposium.

The overview includes contributions of the very early period when interest was focussed mainly on the determination of the physicochemical properties of the fast growing number of newly detected viruses, and of the pioneering period when fundamental discoveries of the nature of viruses were made. The concepts that derived from those studies made the development of modern virology possible. Some highlights of the present period were presented describing the findings of selected virus families. This part was followed by a description of the results which were relevant to problems of how viruses become pathogens, and the role of the immune response to virus infections. Finally, attention was drawn to the contributions of molecular studies which became important not only for the field of virology but also for life sciences in general.

Introduction

A century of virology is cause enough to remember with respect those tasks successfully completed in this field, to recognise ongoing work and to express wishes for success in the future. These aims reflect the spirit that inspired the organisers of this anniversary celebration to deal separately with the contributions of German virologists to the breathtakingly rapid development of our scientific discipline world-wide. This kind of inspiration, however, tends to be difficult to interpret correctly. In light of the international co-operation in science and the corresponding cosmopolitan behaviour of many scientists, it may even be questionable as to whether it is at all feasible to speak truly of national contributions. In the following, I will focus on research that was undertaken in German institutes, even though guest scientists from other countries and all other kinds of international co-operation may have contributed to the work in many cases. However, it will only be possible to recollect a selection of these achievements. These will be limited to work that either resulted in the discovery of previously unknown phenomena, or develop into new ideas, or created new methods that opened the pathways and defined the final objectives which led to the current knowledge and present understanding of our discipline in modern science. Repetitions or mere improvements upon imports of knowledge from other countries will not be considered.

In taking the freedom to express my own personal view on the contributions of German virologists on an international scale and without pretending a historical-biographical professionalism, I fully realise that my selection and evaluation will appear to be subjective. In doing so, I may offend certain individuals and would therefore like to apologise in advance. In any case, I will not sing a hymn of praise but will try to follow the admonishment of Baco von Verulam as quoted by Immanuel Kant in his 'Kritik der reinen Vernunft' (2nd edition): *De nobis ipsis silemus*. This will be somewhat less difficult for me since the topic of veterinary virology will be treated separately by Marian C. Horzinek.

The early post-Loeffler-Frosch period

The history of virology is a particularly good example of how scientific achievements are directly related to the prevailing way of thinking, and to the state of technical development, and to the methods available at that time. It also shows how the past 100 years of virology have forged new concepts and provided new insights into the history of life.

The first era of virology was dominated by bacteriologists or even hygienists who took advantage of the experimental procedures already used so successfully by the discoverers of foot and mouth disease virus. Even at the beginning of the 1950's I was told that a "good bacteriologist is also a good virologist". This attitude therefore clearly shows why virology did not become a separate discipline at German universities until the 1960's.

During that early period the viral aetiology of a large number of infectious diseases was recognised and even viruses which cause tumours were identified. It is remarkable however, that German scientists have taken comparatively little interest in the primary discovery of human pathogenic viruses. Herpes simplex virus (HSV) was among the few exceptions. The unequivocally infectious nature of HSV was recognised 1919 by A. Löwenstein. He demonstrated that virus retrieved from vesicles of herpes labialis produced lesions on the cornea of the rabbit [1]. Forty years later K.E. Schneweis found that 2 serotypes of HSV, HSV-1 and HSV-2, can be differentiated, which are associated with differences in the clinical manifestation of infection [2]. Whereas HSV-1 predominates in infections "above the belt", HSV-2 is associated with genital disease. This discovery breathed new life in herpes virus research. Marburg disease virus, discovered in 1967 by R. Siegert and W. Slenczka [3], was substantially characterised in Marburg and became the first representative member of a new virus family, the *Filoviridae*. In 1979 H. zur Hausen and L. Gissmann [4] discovered in a B-lymphoblastoid cell line derived from an African green monkey a widely distributed B-lymphotropic polyomavirus. In this context Borna disease virus might also be mentioned. It was originally identified by W. Zwick (1926) as a pathogen of horses [5]. However, during the last few years it has attracted much wider attention since growing evidence indicates that Borna virus causes behavioural alterations or psychiatric disorders in humans and animals [6]. Finally, the laboratories of H. zur Hausen and H. Pfister (for [7]) contributed significantly to the world-wide efforts to identify new papillomaviruses. Presently no fewer than 80 different types have been identified, and in addition, more than 50 partial sequences are known, pointing to still more types of such viruses.

During this first period of virology interest focused primarily on the determination of size and shape of the bewildering variety of viruses detected, their sensitivity to chemical and physical agents, their host range and their differences in the manifestation of diseases caused by infection. Application of physicochemical and chemical techniques used in biochemistry and their continuous improvements helped to define the nature of viruses. By ultrafiltration, ultracentrifugation and electron microscopy, developed and applied by H. Bechhold and M. Schlesinger in Frankfurt, and Helmut Ruska in Berlin, in particular, the size and morphology of many viruses was determined. Mrowka, a veterinarian at the former leasehold German naval base at Tsingtao, China, was one of the first to use chemical procedures for the isolation of viruses, as early as 1912. He succeeded in precipitating fowl plague virus from infectious blood serum by means of tannin, without destroying infectivity. He concluded that the virus behaved in all respects like a colloid globulin and should be regarded as such [8]. Twenty years later differential centrifugation and ultrafiltration allowed M. Schlesinger

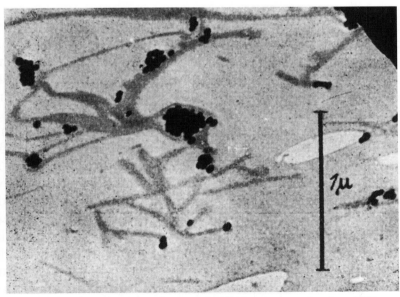

Fig. 1. First electron micrograph of tobacco mosaic virus [10]

Fig. 2. The DNA of bacteriophage T2, liberated from the head of the phage by osmotic shock [12]

(1933) to purify bacteriophage particles in sufficient amounts for various further analyses. He not only obtained important information about the dimensions of such viruses but showed simultaneously that pure phage material consists only of protein and DNA in roughly equal amounts [9]. This led for the first time to the suggestion that viruses in general may be composed of nucleoprotein. In 1939, H. Ruska and co-workers presented the first electron micrograph of any virus, the tobacco mosaic virus (Fig. 1), using a microscope built by his brother Ernst [10]. Two years later he was the first to show how bacteriophages are adsorbed to the surface of their bacterial host [11]. In this context I would like to recall the aesthetic electron micrographs presented in 1962 by A. K. Kleinschmidt and his colleagues [12] which show the DNA molecule of bacteriophage T2 being liberated from the head of the phage particle by osmotic shock, published in several text books (Fig. 2). The Kleinschmidt spreading technique allowed the correct determination of lengths as well as the determination of higher order structures of nucleic acids. Brownian movement brings phage particles into random collision with their host cell which is, as originally described by M. Schlesinger (1932), the first step leading to phage adsorption [13]. In 1954, W. Weidel presented the first evidence for the nature of a corresponding bacterial receptor [14]. G. Koch (1958) characterised as a lysozyme the enzyme responsible for phage release from a bacterial cell by lysis-from-within [15].

From all studies it became clear that viruses are autoreproductive particles ranging in size from the smallest bacteria to the largest known biologically active macromolecules. The most intriguing question remaining at that time was: Do viruses represent the transition from inanimate nature to the typical life? This question, intensively discussed by vitalists as well as by mechanists, became even more accentuated when in 1935 Wendell M. Stanley (Princeton) published the "Isolation of a crystalline protein possessing the properties of tobacco mosaic virus" [16]. However, if we adhere to the principle of the "whig interpretation of history", which evaluates the past on the standard of its significance for the present, today all these questions appear to be of minor interest.

The pioneering period

Immediately realising the utmost significance of Stanley's discovery, Adolf Butenandt, who at that time worked successfully on oestrogen, made a far-reaching decision. Together with F. von Wettstein and A. Kühn in 1938 he established a working group for virus research at the Kaiser-Wilhelm Institut für Biochemie in Berlin-Dahlem. G. Schramm was nominated to head its biochemical section and G. Melchers to be responsible for the genetic part. After the war that working group, which was later joined by H. Friedrich-Freksa, G. Bergold and W. Schäfer, continued with their investigations in Tübingen. The Max-Planck-Institut für Virusforschung, which emerged from this initiative in the 1950's became a focal point for virus research and was prominently involved in the development of molecular biology in Germany. More than that: Tübingen institutes became also the elite school for virology in Germany, which influenced the development of our discipline enormously. Thus, for example, more than 20 of Schäfer's descendants received prominent positions in national and international institutions.

On a par with the establishment of molecular virology in Tübingen, Richard Haas in Freiburg (Fig. 3) put considerable emphasis on medical virology, thereby promoting virology as a new field of research and application in medicine. He really was the forerunner of modern medical virology in Germany. His spirit was carried on by R. Thomssen, who has contributed enormously in tying together medical and molecular virology. He was often ahead of the time, e.g. when he developed the radioimmune assay before the Nobel prize was awarded for this technique [17]. It should also be mentioned that H. J. Eggers, K.-E. Schneweis, and R. Kandolf in particular also played a large part in the bringing together of basic and applied virology. They became particularly known for their work on antiviral agents (for [18]) and on herpes simplex virus pathogenesis (for [19]) or on picornavirus-induced myocarditis (for [20]), respectively.

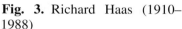

Fig. 3. Richard Haas (1910–1988)

Fig. 4. Gernot Bergold (∗ 1911)

Gernot Bergold (Fig. 4), who left Tübingen in 1948 for a leading position in Canada, can be very rightly regarded as the founder of biochemical insect virology. After a long period of errors in the research on inclusion body diseases of insects, in the 1940's he was able to elucidate the viral aetiology of the polyhedrosis disease of *Bombyx mori* and of another caterpillar disease, the granulosis disease. In both cases, he biochemically characterised the rod-like, DNA-containing viruses and discovered that they were embedded in protective, non-infectious protein structures, the so-called polyhedra [21, 22]. He also showed that infectious virus was released from polyhedra by treatment with diluted alkaline solution (Fig. 5).

Plant viruses, in particular tobacco mosaic virus (TMV), proved to be suitable as a model to elucidate the structural properties of viruses, since they could be obtained quite easily. It was found that up to 90% of the protein present in infected plant juice might consist of TMV and reliable methods to quantify virus particles then became available. Gerhard Schramm (Fig. 6), in Berlin, had already detected that treatment with slightly alkaline solution caused TMV to dissociate into subunits with defined size and shape. The isolated subunits could be reaggregated to TMV-like rods, while infectivity was lost [23]. The amino acid sequence of TMV protein later was resolved by A. Anderer (1960), as the first primary structure of any viral protein [24]. Subsequent determination of the protein sequences from different TMV strains and mutants helped H. G. Wittmann (1962) to contribute to the codon assignment for the genetic code, which of course added evidence to the universality of the genetic code [25]. Of exceptional importance was the finding by A. Gierer and G. Schramm [26] in 1956 that the genetic information of TMV resides in its RNA. From this discovery a most important conclusion was drawn that RNA could also be genetic material, a property previously thought to

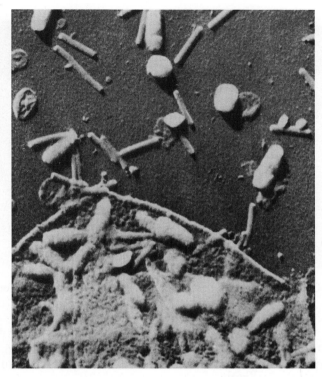

Fig. 5. Polyhedra obtained from *Lymantriz dispar*, dissolved with alkaline solution (Photograph by G. Bergold)

be restricted to just DNA. The possibility of isolating biologically intact RNA by the phenol method has contributed enormously to many facets of molecular biology. A modification described by E. Wecker [27], the "hot phenol method", also allowed the extraction of infectious RNA from enveloped positive stranded viruses. The analytical studies on TMV-RNA by H. Schuster provided the basis for an elucidation of the mechanisms of mutagenicity caused by nitrous acid and hydroxylamine treatments [28, 29]. Based on these results in 1958 A. Gierer and K. W. Mundry succeeded for the first time in generating specific virus mutants [30]. Treatment with chemical mutagens enhanced the mutation frequency, which became a useful tool for genetic studies in general. In 1963 Anderer was the first to demonstrate that an isolated hexapeptide of the TMV-protein forms the minimal structure for an epitope capable of inducing virus-specific antibodies [31].

Werner Schäfer (Fig. 7), the successor to Bergold in the field of animal virology in Tübingen, became acknowledged world-wide for his studies on fowl plague virus (FPV), Newcastle disease virus and encephalomyocarditis virus, as well as RNA tumour viruses. FPV proved to be an excellent paradigm to study structural and functional relationships of enveloped viruses and served as a feasible agent for tracing virus replication, particularly of orthomyxoviruses. Without any doubt FPV was for a long time one of the best known animal viruses, with respect to its physical, chemical, architectural and biological properties [32]. If

Fig. 6. Gerhard Schramm (1910–   **Fig. 7.** Werner Schäfer (∗ 1912)
1969)

one is tempted to give testimony to a fair spirit of competition in the work of
G. Schramm in Tübingen and H. Fraenkel-Conrat in Berkeley, one may also
similarly recognise a competitive parallelism in the way Schäfer and Leslie Hoyle
(Northampton) dealt with FPV and human influenza viruses, respectively. This
became particularly evident when Schäfer found in 1955 that FPV is in fact an
influenza virus [33], and that it might, perhaps through a process of recombination,
exchange host specificity with other influenza A viruses which might contribute
to the frequent occurrence of previously unencountered strains. We know today
that this assumption was close to reality. The model of influenza virus structure
developed by Schäfer showed a filamentous ribonucleoprotein surrounded by a
lipid-containing envelope, into which a glycoprotein, the haemagglutinin (HA)
is incorporated. The HA serves as a ligand during adsorption of the virus to a
cellular receptor, the determinant of which was identified by E. Klenk (1955)
as neuraminic acid [34]. In addition, the HA turned out to be the immunogen
which induces the production of protective neutralising antibodies in the infected
host [35]. Schäfer's proposal to use only the immunogenic glycoprotein for the
vaccine production has meanwhile been realised via subunit vaccines also used
for immunisation against other virus infections. Worth mentioning are the results
obtained by the Tübingen group on the participation of the cell nucleus in the
replication of influenza viruses [36], first indication that the virus envelope is a
virus-specific altered host cell membrane [37], the first indication of the segmented
nature of influenza virus RNA [38] and – already largely forgotten – the first
evidence that the production of viral proteins is possible in subcellular fractions,
i.e. in an in vitro system without employing intact cells [39].

Schäfer's scientific descendants in Giessen later extended the knowledge
about structure and biology of orthomyxo- and parainfluenza viruses, when the

arsenal of methods had been expanded and refined. Recognition of the exceptional segmented structure of influenza viral RNA allowed new insights into viral genetics, into the emergence of new influenza viruses, and into molecular epidemiology (for [40]). Certainly, the results obtained by the Giessen team (mainly H. Becht, W. Garten, H.-D. Klenk, M. Orlich, R. Rott, M. F. G. Schmidt, C. Scholtissek and R. T. Schwarz) on structure, production and biological properties of influenza and parainfluenza viral glycoproteins have set a precedent for subsequent investigations with other viruses (for [41]). This includes post-translational modification of the glycoproteins by the different steps of glycosylation, by employment of new glycosylation inhibitors (for [42]), by palmitoylation and myristoylation (for [43]), and by proteolytic cleavage [44]. In this way the dominant role of these glycoproteins in the initial process of viral replication and their significance as determinants for pathogenicity have been resolved. Though the presence of receptor destroying enzyme of influenza C virus was demonstrated already in 1950, it was characterised only in 1985, by G. Herrler, as a neuraminate-9-O-acetyl esterase [45].

In 1953 Arnold Graffi isolated in Berlin-Buch the causative virus of murine myeloid leukemia of mice [46], named after him the "Graffi virus", which he identified later as a type D retrovirus. It was again W. Schäfer, who, together with Heinz Bauer, introduced basic retrovirus research in Germany. Following the previous experience with the myxoviruses that elucidation of the correlation between structure and function will yield the deepest insight into the nature of viruses, their groups made important contributions to retrovirus research. Characterisation of the different structural compounds of murine and chicken oncornaviruses was without doubt among the highlights of the diverse studies performed in Tübingen and later by several other groups in Germany. The fundamental insights achieved led to the world-wide understanding of the structure of these viruses, and the production of globally employed monospecific antibodies, some of which have been suggested for use in tumour therapy [47]. Completion of our knowledge on the action of the enzyme reverse transcriptase came from Karin Mölling (1971), who discovered RNase H activity and the mechanism of its function as a processively acting exonuclease [48].

The present period

Since the beginning of the 1960s the establishment of virology as a separate discipline at German universities, the possibility of study periods abroad, the continuous development and application of new techniques, but also the frequent use of viruses to study general biochemical and molecular biological aspects have all contributed late, but not too late, to the boost in virology in Germany.

It is interesting to note that in the early 1960's several virologists held the view that the golden age of virology was already over. With all major foundations of molecular biology elucidated no more spectacular results were expected; apparently the "eighth day of Creation" came to an end. Of course, this assumption turned out to be inaccurate. Even though no Nobel-prize awarded discoveries were made in virology in Germany, a number of impressive results have significantly contributed to the mosaic of our current knowledge of the nature of viruses and of their properties as causative agents of infectious diseases. Since a large number of these *tesserae* should be common knowledge, I shall only expand upon a few areas in which German virologists have substantially contributed.

Viroids

Certainly, one of the most remarkable discoveries in plant virology in the post-Schramm era was the simultaneous and independent finding by Theodor O. Diener in the USA and Heinz L. Sänger in Giessen of "naked" small RNA molecules as a new kind of autonomously replicating subviral plant pathogens known today as viroids. Previously Sänger had successfully studied the structural and genetic interactions of the two particles of bipartite tobacco rattle virus whose unique helper mechanism he could elegantly explain [49]. Based on this experience he characterised the causative agent of exocortis disease of citrus as a viroid [50]. He then succeeded in isolating and purifying several other viroids, resulting in a detailed biochemical, physicochemical and morphological characterisation. Thus, in collaboration with G. Klotz, D. Riesner, H. J. Gross and A. K. Kleinschmidt [51] he was able to demonstrate in 1976 that "viroids are single-stranded covalently closed circular RNA molecules existing as highly base-paired rod-like structures" with a molecular weight of 120,000 corresponding to ca. 360 nucleotides. In 1978 both the nucleotide sequence and secondary structure of the first viroid RNA was published [52]. His subsequent studies later undertaken in Martinsried on the relation between viroids structure and function and on viroid replication rendered viroids the best characterised class of small RNA molecules next to tRNAs.

Hepadnavirus

It is generally agreed that German virologists around H. Schaller, P.-H. Hofschneider, W. Gerlich, and H. Will, contributed enormously to our current knowledge on hepadnaviruses, particularly hepatitis B virus (HBV). Schaller and Hofschneider were involved in cloning and sequencing the whole HBV-genome [53, 54], through which it became possible not only to understand this virus' structure but eventually also to produce the first anti-cancer vaccine. It was H. Will in Schaller's laboratory who obtained the first cloned infectious DNA [55]. Characterisation of viral transcripts by Schaller's group and study of the viral DNA polymerase revealed the full replication strategy of the hepadnaviral genome [56]. W. Gerlich deserves credit for elucidating the structure-function relationship of HB-S and HBe proteins [57, 58]. Hofschneider's group showed that the HBx protein acts as a transactivator, stimulating a striking variety of promoters, which do not share any known cis-regulator element [59]. This group also showed that HBx is frequently present in liver carcinomas. Gerlich's group demonstrated that HBx is in fact able to transform immortalised hepatocytes [60]. Most surprising was the observation by Hofschneider that the pre-S-domain of the HBV genome also possesses a transactivating effect [61]. Finally, some indications on the pathogenesis of HBV-infection was obtained with virus variants isolated by H. Will.

Papillomaviruses

Since the beginning of this century, viruses had been known to be the causative agent of human skin warts, genital warts and laryngeal papillomas. For decades, wart viruses had been barely characterised due to the lack of in vitro systems for viral propagation and it was generally believed that there would be only a single type of human papilloma virus. Stimulated by the tendency of certain types of human warts to malignant conversion, H. zur Hausen, L. Gissmann, and H. Pfister started a systematic analysis of virus isolates from individual warts in the mid-1970's and soon established the heterogeneity of papilloma viruses by characterising HPV1 and HPV4. With the advent of recombinant DNA technology, these investigations led to cloning and characterisation of papillomaviruses from different sources. For instance HPV6 and HPV11 from *condylomata acuminata* and laryngeal papillomas, HPV8, 19, 20, and 25 from patients with *epidemodysplasia verruciformis*, viruses known to correlate with increased risk of developing skin cancer, HPV13 from Heck's disease of the oral mucosa, and HPV16 and 18 from cervical cancers. HPV 16 or 18 can be detected in up to 70% of carcinomas of the cervix uteri, and both are now recognised by the WHO as the major cause of cervical cancer [62–65].

Seroepidemiological studies in Pfister's laboratory during the 1980's indicated that HPVs, originally assumed to be restricted to patients with *epidermodysplasia verruciformis* (EV), are widespread in the general population. This was most recently confirmed by the demonstration – in plucked hairs – of EV-virus-specific and related HPV DNA sequences in a considerable proportion of asymptomatic controls. Such sequences were similarly found in more than 50% of cutaneous squamous cell carcinomas in the general population [65].

In the past decade zur Hausen's group discovered a number of intracellular and intercellular signalling pathways that regulate cell differentiation but that also influence HPV oncogene activity [66]. Similarly, Pfister and colleagues identified the cellular transcription factor YY1 as a repressor of HPV16 oncogene transcription, and showed frequent deletion of YY1 binding sites from extrachromosomal HPV16 DNA within cervical cancers [67]. This likely leads to increased activity of the oncogene promoter and suggests another important step in tumour progression.

Viral oncogenes

Germans were involved in other innovative studies on viral oncogenes. T. Graf and H. Beug [68] are particularly known as the discoverer of the retroviral erb B oncogene. K. Mölling [69, 70] found the first retroviral oncogene products, Myc and Myb, located in the cell nucleus, and their DNA-binding ability in vitro. She also discovered the first serine/threonine protein kinase encoded by the oncogene mil/raf [71]. B. Fleckenstein in co-operation with W. Haseltine, Boston, identified the tax-gene product of human T-cell leukaemia virus type 1 as the T-cell transforming protein [72]. In his highly acknowledged studies on herpesvirus saimiri, Fleckenstein described new transforming genes. Thus, in a subgroup A strain an oncogene, stpA, which is responsible for peripheral T-cell lymphomas in transgenic mice, was detected, mapped and characterised [73]. At the homologous position in the genome of a subgroup C-strain the information is localised for two oncogenes stpC and tip. The first strongly transforms rodent fibroblasts while the product of tip interacts specifically with T-cell specific tyrosine kinase Lck, which might explain the T-cell tropism of transformation by herpesvirus saimiri [74].

Finally I emphasise Wolfgang Deppert's analysis of the interaction of the SV40 T antigen with the cellular regulator protein p53 [75]. The p53 protein is the most famous protein in tumour biology, as it is a tumour suppressor whose gene is genetically altered in about 50 to 60% of all human cancers. Deppert's finding that p53 exhibits $3'$- to $5'$-exonuclease activity substantially extended our view concerning its role as a "guardian of the genome" such as control of homologous recombination and the possibility that p53 might act as an external proof-reader for polymerase alpha in SV40 DNA replication [76].

Pathogenesis of virus infections

During the last 30 years, we have obtained more and more results that are relevant to the questions as to how viruses become pathogens. First demonstrated with influenza viruses and then confirmed for an increasing number of other viruses, pathogenicity is of polygenic nature. However, in addition to the necessity of an optimal gene constellation [77] the Giessen virologists demonstrated the importance of the structure of the cleavage site of the HA glycoprotein of influenza viruses and the F protein of parainfluenza viruses in determining pathogenicity of these viruses, and also pointed to the potential of the proteases secreted from co-infecting bacteria for enhancing viral invasiveness (for [78]). There is now evidence for an analogous effect with *Filoviridae* as shown by Heinz Feldmann in Marburg [79].

The Würzburg group, in cooperation with Martin Billeter from Zürich, has shown that in infected brain cells from patients with subacute sclerosing panencephalitis (SSPE) induced by measles virus the viral envelope glycoproteins are markedly underexpressed or even absent. This is apparently caused by the presence of a mutated stop codon in the corresponding genes. In addition, in measles virus cloned from infected brain tissue, a biased hypermutation has been demonstrated in the M gene, which leads to an exchange of up to 50% of a particular C residue to U, possibly caused by the action of a cellular duplex RNA-dependent adenosine deaminase activity found in human neural cell extracts [80, 81]. Thus, measles virus formation in brain cells seems to be associated with an abrogation of M protein function, as has also been suggested for abortive infection of influenza virus in brain cells [82].

In some virus-induced diseases of the central system (CNS), the lesions very much resemble the neuropathological changes observed in experimental allergic encephalitis. Based on these observations, the virologists around V. ter Meulen and H. Wege in Würzburg have established two interesting animal models in which a coronavirus or a measles virus infection leads to an autoimmune inflammatory disease process in the CNS. Both virus infections induce the activation of CD4+ T-cells against brain specific antigens, which become perpetuated after virus replication has ceased. Similarly, as shown by the Giessen group, vesicular stomatitis virus, when grown in brain cells, causes demyelination, too. In this case, myelin basic protein was found incorporated into the envelope during virus maturation [83–85]. These results suggest that unmasking of CNS membrane components and/or incorporation of host-specific antigens into the viral envelope and subsequent priming of self-reactive immune response might be a common pathogenic mechanism underlying the post-infectious encephalitis syndrome as already hypothesised in 1969 [86].

Otto Haller, when coming from Zürich to Freiburg, continued his studies on the Mx family of interferon-induced antiviral proteins, particularly the human MxA protein. Investigations on MxA transgenic mice have shown that MxA has a powerful antiviral effect also in vivo. In the Thogoto virus model he demonstrated for the first time a mechanism by which MxA exerts its protective activity: MxA binds to the incoming viral RNP in the cytoplasm of infected cells, thus preventing its import into the nucleus and consequently viral genome amplification and transcription [86a].

Virus interaction with the immune system

In the mid 1970's U. Koszinowski and R. Thomssen reported on lysis mediated by T-cells and restricted by H-2 antigen of target cells infected with vaccinia virus [87]. This was the first virological confirmation of the fundamental work on MHC restriction of virus-specific T-cells published only shortly before by Zinkernagel and Doherty. Furthermore in determining the requirements for generation of virus-specific cytotoxic T-cells, Koszinowski [88] and others [89] found for example, that fusion of Sendai or fowl plague virus with target cell membranes is required for T-cell recognition.

It is due to Fritz Lehmann-Grube that attention was drawn to the role of T-cell mediated cytotoxicity in the elimination of viruses in the infected organism. He contributed a great deal to our understanding of the mechanism of the immunopathogenesis of lymphocytic choriomeingitis of mice. This disease, which was originally studied by Erich Traub, became the paradigm for virus diseases, in which the infecting virus by itself does not affect vital functions but the outcome of the disease is caused by T-cell-dependent immunopathological reaction [90, 91]. A similar mechanism of immunopathogenesis was found underlying Borna disease [92] by a research group in Giessen and hepatitis A by A. Vallbracht and B. Fleischer [93].

Viruses are true survival artists and have invented different tricks to escape the immunological defence. Suppression of the host immune system was first documented 90 years ago by the German paediatrician Clement von Pirquet who observed that the tuberculin skin test of immune individuals was depressed during the course of acute measles virus (MV) infection [94]. A breakthrough in the understanding of this important phenomenon, regarded as a major cause of the high mortality of MV infection, came from V. ter Meulen's group [95]. They convincingly demonstrated that – in direct contrast to commonly held opinion – responding lymphocytes do not themselves need to be infected in order to be suppressed but rather that the contact with both viral glycoproteins triggers immune suppression.

Cytomegaloviruses (CMV) manipulate the immune system on several levels. U. Koszinowski and his co-workers have delivered important contributions on the disturbance of formation and transport of MHC molecules in CMV-infected cells, which prevent or reduce their expression on the cell surface. Recently, they found three new CMV proteins that interact with this process (for review see [96]).

Contributions to molecular biological studies

Obviously World War II and restrictions imposed by the Allies on particular fields of research, such as genetics, prevented German scientists in the post-war period to participate in development of molecular genetics as initiated by the "phage group" around Luria, Delbrück, Hershey and others. Unfortunately, German scientists also missed the boat in the beginning of recombinant DNA revolution where viruses again played a central role. Nevertheless, a few impressive contributions in the further development of that area which became important not only for molecular virological studies, but also for life sciences in general are mentioned here. Thus, I would like you to remember that R. Jaenisch was definitely the first to produce transgenic animals. In 1975, while still in Hamburg, he was able to show that after infection of early mouse embryos with Moloney leukaemia virus the viral genome became incorporated into cells to the germ line and that the integrated genetic information was inherited in accordance with Mendel's rules [97]. He found eight years later that the integration can lead to recessive lethal mutation of the cellular gene which carries the provirus. With this discovery Jaenisch was also the first who described the phenomenon of insertion mutagenesis in mammals (for [98]). In 1960 Hofschneider isolated infectious DNA from phages for the first time [99] and described in 1974 the isolation and properties of the replicative form of phage M12 RNA which is relevant for the replication of many RNA viruses [100]. With phage ϕX174 first evidence was obtained for genetic recombination of single stranded DNA. Once suitable selective genetic markers had been developed, D. Pfeifer (1961) found that recombination could be detected at a level of 10^{-4} and 10^{-5} recombinants per progeny virus [101]. It was much later in Giessen that non-homologous RNA recombination was detected with influenza viruses [102]. In the field of virus evolution the group of Manfred Eigen, who developed the "quasispecies" concept, has made important contributions concerning the experimental coupling of mutation and selection [103]. As early as the 1970's M. Sumper, a member of Eigen's group, had shown that genomic RNA could be recognised and reproduced by Qß-replicase, the RNA polymerase of phage Qß, and that during replication under certain environmental conditions defined phenotypical properties may be selected [104]. H. Schaller studied the prokaryotic promoter structure in 1977 and defined the minus ten region, the "Schaller-box", as the polymerase binding site [105]. This was exemplified on phage fd, the first known filamentous, circular single-stranded DNA Ff phage, isolated by Hoffmann-Berling (1961). In 1977 J. Messing introduced part of the lac regulatory region into the genome of M13 [106], another Ff phage, which converted this phage to a most suitable cloning vehicle and made Sanger-sequencing with this single-stranded DNA vector genome quite easy.

M13 vectors soon became the most important vehicles for shotgun cloning and sequencing, and continue to be so today. It should also be mentioned that the first information about the significance of the baculovirus vector came from Giessen [107]. The strong promoter of cytomegalovirus is also widely used as the driving force in eukaryotic expression constructs in several aspects of gene technology. It is, as found in 1985 by B. Fleckenstein's team constitutively active and is not controlled by transactivating or other viral factors but can be regulated by cellular transcription factors [108]. In 1978 G. Hobom was the first to describe structure and function of the bacteriophage lambda origin of replication [109]. H. Lehrrach and A. Frischauf developed from the phage lambda the so-called EMBL phages [110], which proved to be a most suitable basis for the construction of gene banks.

Reverse genetics was extended to negative stranded RNA viruses by K.-K. Conzelmann (1994). He succeeded in molecular cloning an infectious cDNA of rabies virus, which proved useful as a new vector system [111]. Hobom (1994) was able to construct a cDNA system for in vivo expression of the segmented influenza viral RNA by RNA polymerase I [112]. Cloning the whole, up to 230 Kbp containing infectious genomes of herpes viruses by U. Koszinowski and W. Hammerschmidt, promises important results for these viruses in the future [113, 114].

In a follow-up study of his discovery that adenovirus 12 (Ad12) DNA persists in transformed hamster cells in an integrated state, Walter Doerfler (1978) found that integrated Ad12 DNA becomes modified by methylation and that integration also changes the methylation pattern of cellular DNA sequences (for [115]). These observations stimulated further studies on the role of DNA methylation in the regulation of eukaryotic gene expression.

Epilogue

It is certainly possible to criticise concerning the development of virology in Germany and also to point out reasons as to why our research may perhaps have had certain shortcomings, when compared to research carried out in some other countries. I must admit, however, that what had seemed first like a major burden to me, then gained a touch of personal pride even after realizing that not only it would be impossible to mention every important contribution from our country to virus research but it would also be impossible to do justice to these contributions. The chosen examples – personally biased – can therefore neither be regarded a complete nor a truly representative selection of all innovative virus research carried out in Germany over the last century. Many of the studies not mentioned here, have made equally reputable contributions to international virology.

This is particularly true concerning clinical virology. On a daily basis, fundamental discoveries are rare events. Their true significance comes to light when sensitive methods, which in part have been developed by clinical virologists, are applied and as consequence, precise and definitive results are obtained, enabling a diagnosis to be made and a clinical alarm or all-clear to be sounded, or when epidemiological relationships can be established and infection chains thus uncovered. Moreover, through the development and testing of vaccines and methods for virus inactivation and also in the evaluation of the effect of chemotherapeutics – albeit to-date not yet as successful – clinical virology has proven itself to be invaluable. It is obvious that some areas of basic virological research have their roots in clinical observation. Furthermore, although German clinical virologists often tend not to make headline news, new discoveries they have made are highly regarded by their international community. For example, I might recall the standardisation of diagnostic methods or the ease with which new findings have repeatedly been qickly introduced into general praxis. The *Deutsche Vereinigung zur Bekämpfung der Viruskrankheiten* has no doubt played a significant role in this. Ultimately as a result of these successes, virology's reputation has not only been boosted in the eyes of the general public but more importantly it is also regarded in a different light by those institutions who provide substantial support for research. Although I could recount the names of many noteworthy clinically orientated virologists, I do not believe that I am wrong in choosing Gisela Ruckle-Enders from Stuttgart as an example. With a background in basic research, she earned special recognition in the area of epidemiology of intrauterine and perinatal virus infections whilst running a virus-diagnostic laboratory in a truly exemplary manner. This kind of fruitful juxtaposition of theory and practice or of more basic and more applied research, something that can also be seen in the transdisciplinary makeup of the *Gesellschaft für Virologie*, will become even more important in the future. This is especially evident if we think of the origins of virology and thereby address the questions, which again increasingly come to the forefront, concerning the mechanisms by which viruses become pathogenic agents and with which means we can better confront the problems of virus infection.

Virology is sometimes regarded as one of the jewels of German research. All the same, it would be dangerous, even on this 100th anniversary, to be too high handed in this regard in such a review. In recollecting such contributions I hope that our young adept scientists will be guided by the desire to equal the achievements made by their predecessors, even to outdo them in the future.

References

1. Löwenstein A (1919) Mün Med Wochenschr 66: 769
2. Schneweis KE (1962) Z Immunforsch 124: 24
3. Siegert R, Shu H-L, Slenczka W, Peters D, Müller G (1967) Dtsch Med Wochenschr 51: 2341
4. zur Hausen, Gissmann L (1979) Med Microbiol Immunol 167: 137
5. Zwick W, Seifried O, Witte J (1926) Z Inf Krankh Haustiere 30: 42
6. Rott, R, Herzog S, Bechter K, Frese K (1991) Arch Virol 118: 295
7. zur Hausen H, de Villiers EM (1994) Ann Rev Microbiol 48: 427
8. Mrowka (1912) Zbl Bakt Parasitkd 67: 249
9. Schlesinger M (1933) Biochem Z 273: 306
10. Kausche GA, Pfankuch E, Ruska H (1939) Naturwissenschoften 27: 292
11. Ruska H (1941) Arch Virusforsch 2: 345
12. Kleinschmidt AK, Lang D, Jacherts D, Zahn RK (1962) Biochim Biophys Acta 61: 131
13. Schlesinger M (1932) Z Hyg Infektkrankh 114: 149
14. Weidel W (1958) Ann Rev Microbiol 12: 27
15. Koch G, Dreyer WJ (1958) Virology 6: 291
16. Stanley WM (1935) Science 81: 644
17. Thomssen R (1963) Nature 198: 613
18. Eggers HJ (1982) In: Handbook of Experimental Pharmacology 61: 377
19. Schneweis KE, Brado M, Ebers B, Friedrich A, Olbrich M, Schüler W (1988) Med Microbiol Immunol 177: 1
20. Kandolf R (1993) Intervirology 35: 140
21. Bergold G (1947) Z Naturforsch 2b: 122
22. Bergold G (1948) Z Naturforsch 3b: 338
23. Schramm G (1943) Naturwissenschaften 31: 94
24. Anderer FA, Uhlig H, Weber E, Schramm G (1960) Nature 186: 922
25. Wittmann HG (1962) Z Vererbungslehre 93: 491
26. Gierer A, Schramm G (1956) Z Naturforsch 11b: 138
27. Wecker E (1959) Virology 7: 241
28. Schuster H (1960) Biochem Biophys Res Commun 2: 320
29. Schuster H (1961) J Mol Biol 3: 447
30. Gierer A, Mundry KW (1958) Nature 182: 1457
31. Anderer FA (1963) Biochim Biophys Acta 71: 246
32. Schäfer W (1963) Bacteriol Rev 27: 1
33. Schäfer W (1955) Z Naturforsch 10b: 81
34. Klenk E, Faillard H, Lempfried H (1995) Z Physiol Chem 301: 235
35. Schäfer W, Zillig W (1954) Z Naturforsch 9b: 779
36. Breitenfeld PM, Schäfer W (1957) Virology 4:328
37. Wecker E (1957) Z Naturforsch 12b: 208
38. Scholtissek C, Rott R (1963) Nature 199: 200
39. Mueller GC, v Zahn-Ullmann S, Schäfer W (1960) J Biol Chem 235: 660
40. Rott R (1997) Berl Münch Tierärztl Wochenschr 110: 241
41. Klenk H-D, Garten W (1994) Trends Microbiol 2: 39
42. Schwarz RT, Datema R (1980) Trends Biochem Sci 5: 65
43. Veit M, Schmidt MFG (1998) Methods Mol Biol 88: 227
44. Klenk HD, Rott R, Orlich M, Blödorn J (1975) Virology 68: 426
45. Herrler G, Rott R, Klenk H-D, Müller H-P, Schukla AK, Schauer R (1985) EMBO J 4: 1503

46. Graffi A, Bielka H, Frey F, Scharsch F, Weiss R (1955) Wien Klin Wochenschr 105: 61
47. Schäfer W, Bolognesi DP, de Noronha F, Fischinger PJ, Hunsmann G, Ihle JN, Moennig V, Schwarz H, Thiel H-J (1977) Med Microbiol Immunol 164: 217
48. Mölling K, Bolognesi DP, Bauer H, Busen W, Plassmann HW, Hausen P (1971) Nature New Biol 234: 240
49. Sänger HL (1969) J Virol 3: 304
50. Sänger HL (1972) Adv Biosci 8: 103
51. Sänger HL, Klotz G, Riesner D, Gross HJ, Kleinschmidt AK (1976) Proc Natl Acad Sci USA 73: 3852
52. Domdey H, Jank P, Sänger HL, Gross HJ (1978) Nucleic Acids Res 5: 1221
53. Burell CJ, Mackay P, Greenaway PJ, Hofschneider HP, Murray K (1979) Nature 279: 43
54. Pasak M, Goto T, Gilbert W, Zink B, Schaller H et al. (1979) Nature 282: 575
55. Will H, Cattaneo R, Koch HG, Darai G, Schaller H et al. (1982) Nature 299: 740
56. Cattaneo R, Will H, Schaller H (1984) EMBO J 3: 2 191
57. Heermann KH, Goldmann U, Schwartz W et al. (1984) J Virol 52: 396
58. Uy A, Bruss V, Gerlich WH, Köchel HG, Thomssen R (1986) Virology 155: 89
59. Zahn P, Hofschneider PH, Koshy R (1988) Oncogene 3: 169
60. Höhne M, Schaefer S, Seifer M et al. (1990) EMBO J 9: 1 137
61. Schlüter V, Meyer M, Hofschneider PH et al. (1994) Oncogene 9: 3 335
62. Gissmann L, zur Hausen H (1976) Proc Natl Acad Sci USA 73: 1 310
63. Gissman L, Pfister H, zur Hausen H (1976) Virology 76: 569
64. Durst M, Gissmann L, Kenberg H, zur Hausen H (1983) Proc Natl Acad Sci USA 80: 3 812
65. Pfister H (1992) Semin Cancer Biol 3: 263
66. zur Hausen H (1994) Curr Topics Microbiol Immunol 186: 131
67. Pfister H (1996) Obstet Gynecol Clinics 23: 579
68. Graf T, Beug H (1983) Cell 34: 7
69. Donner P, Greiser-Wilke I, Moelling K (1982) Nature 296: 262
70. Donner P, Bunte T, Greiser-Wilke I, Moelling K (1983) Proc Natl Acad Sci USA 80: 2 861
71. Moelling K, Heimann, B, Beimling P, Rapp UR, Sander T (1984) Nature 312: 558
72. Grassmann R, Dengler C, Müller-Fleckenstein I et al. (1989) Proc Natl Acad Sci USA 86: 3 351
73. Jung JU, Trimble JJ, King NW et al. (1991) Proc Natl Acad Sci USA 88: 7 051
74. Biesinger B, Tsygankov A, Fickenscher H et al. (1995) J Biol Chem 270: 4 729
75. Wiesmüller C, Cammenga J, Deppert W (1986) J Virol 70: 737
76. Mummenbrauer T, Janus F, Müller B et al. (1996) Cell 85: 1 089
77. Rott R, Orlich M, Scholtissek C (1979) J Gen Virol 44: 471
78. Rott R, Klenk H-D, Nagay Y, Tashiro M (1995) Am J Resp Crit Care Med 152: 516
79. Volchkov VE, Feldmann H, Volchkova VA, Klenk H-D (1998) Proc Natl Acad Sci USA 95: 5 762
80. Liebert VG, Baczko K, Budka H, ter Meulen V (1986) J Gen Virol 67: 2 435
81. Cattaneo R, Schmidt A, Spielhofer P et al. (1989) Virology 173: 415
82. Schlesinger RW, Bradshaw GL, Barbone F et al (1989) J Virol 63: 1 695
83. Watanabe R, Wege H, ter Meulen V (1983) Nature 305: 150
84. Liebert VG, Lington, CC, ter Meulen V (1988) J Neuroimmunol 29: 139
85. Rott O, Herzog S, Cash E (1994) Med Microbiol Immunol 183: 195
86. Drzeniek R, Rott R (1969) Arch Allergy 36: 146

86a. Kochs G, Trost M, Janzen G, Haller O (1998) Methods: A companion of Methods in Enzymology 15: 255

87. Koszinowski U, Thomssen R (1975) Eur J Immunol 5: 246

88. Koszinowski UH, Gething AH, Waterfield MD, Klenk HD (1980) J Exp Med 151: 945

89. Kurrle R, Wagner H, Röllinghoff M, Rott R (1979) Eur J Immunol 9: 107

90. Lehmann-Grube F (1984) Intervirology 22: 121

91. Lehmann-Grube F, Moskophidis D, Löhler J (1988) Ann NY Acad Sci 532: 238

92. Narayan O, Herzog S, Frese K, Rott R (1983) Science 220: 1 401

93. Vallbracht A, Maier K, Stierhof YD et al. (1989) J Infect Dis 160: 209

94. Pirquet Cv (1908) Dtsch Med Wochenschr 30: 1 297

95. Schlender J, Schnorr J-J, Spielhofer P et al. (1996) Proc Natl Acad Sci USA 93: 13 194

96. Hengel H, Brune W, Koszinowski U (1998) Trends Microbiol 6: 190

97. Jaenisch R, Fan H, Croker B (1975) Proc Natl Acad Sci USA 72: 4 008

98. Jaenisch R (1982) Hoppe-Seyer's Z Physiol Chem 363: 1 267

99. Hofschneider PH (1960) Z Naturforsch 15b: 441

100. Amann J, Delius H, Hofschneider PH (1964) J Mol Biol 10: 557

101. Pfeifer D (1961) Z Vererbungslehre 92: 312

102. Khatchikian D, Orlich, M, Rott R (1989) Nature 340: 156

103. Eigen M, McCaskill J, Schuster P (1983) J Physiol Chem 92: 6 881

104. Sumper M, Luce R (1975) Proc Natl Acad Sci USA 72: 162

105. Schaller H, Gray C, Herrmann K (1977) Proc Natl Acad Sci USA 74: 737

106. Messing J, Gronenborn B, Müller-Hill B, Hofschneider PH (1977) Proc Natl Acad Sci USA 74: 3 642

107. Kuroda K, Hauser C, Rott R, Klenk H-D, Doerfler W (1986) EMBO J 5: 1 359

108. Boshart M, Weber F, Jahn G et al. (1985) Cell 41: 521

109. Grosschedl R, Hobom G (1979) Nature 277: 621

110. Lehrrach H, Frischauf A (1983) J Mol Biol 170: 827

111. Schnell MJ, Mebastion T, Conzelmann K-K (1994) EMBO J 13: 4 195

112. Neumann G, Zobel A, Hobom G (1994) Virology 202: 477

113. Messerle M, Crnkovic I, Hammerschmidt W et al. (1997) Proc Natl Acad Sci USA 94: 14 759

114. Decluse H, Hilsenberger T, Pich D et al (1998) Proc Natl Acad Sci USA 95: 8 245

115. Sutter D, Westphal M, Doerfler W (1978) Ann Rev Biochem 52: 93

Authors' address: Dr. R. Rott, Justus-Liebig-Universität Giessen, Institut für Virologie, Frankfurter Straße 107, D-35392 Giessen, Federal Republic of Germany.

Importance and impact of veterinary virology in Germany

M. C. Horzinek

Veterinary Faculty, Virology Unit, Utrecht University, Utrecht, The Netherlands

Summary. The causative agent of tobacco mosaic and of foot and mouth disease (FMD) were recognized in 1898 as "filterable" or "invisible" – and eventually termed "virus". Four years later the viral aetiology of yellow fever was established, and the new discipline took off. Thus animal virology started with a veterinary problem, and Germany's contribution during the following decades came mainly from the chairs of veterinary teaching and research establishments in Giessen, Munich and Hanover, the Riems Institute, and the Federal Research Institute for Animal Virus Diseases in Tübingen. From a superficial bibliometric analysis, a wide divergence in impact figures is noted, with excellent contributions in international virology journals and lesser papers in German veterinary journals. The publications in the observed time frame reveal a fascination by virion structure, physical characteristics and structure-function relationships with little work published in journals dedicated to immunology and pathogenesis.

Scientific priority

The first German connection with virology – though not in the animal field – predates Martinus Willem Beijerinck's historic definition (1898) of the *contagium vivum fluidum:* Adolf Mayer (1843–1942), a chemist from Heidelberg, was appointed at the Agricultural School in Wageningen, the Netherlands, in 1876. He first reported on a disease in tobacco plants in 1882, named it 'tobacco mosaic', and showed that it could be serially transmitted in the apparent absence of microorganisms [6]. The causative agent of tobacco mosaic was to become the first model virus that revealed many secrets of virion structure. Dmitri Ivanovsky (1864–1920) is quoted for his classical filtration experiments in which he demonstrated passage of the causative agent of tobacco mosaic through the pores of a bacteria-proof Chamberland filter. His paper, read before the Academy of Sciences in St. Petersburg, Russia in 1892, is undoubtedly a landmark in the history of virology. Of special significance for interpreting the author's ideas, however, is his dissertation published in German while he was working in Warsaw [5]. In it he reiterated that he was dealing with a microbe, which might have passed the pores of the bacteria-proof filter or might have produced a filterable toxin.

In analysing priority claims one should appreciate conceptual originality rather than comparing publication dates. The polemics surrounding such claims reflect the Olympic spirit in science – *citior, altior, fortior* – giving the illusion that fame can be quantitated. Beijerinck's achievements for virology are sometimes disputed in this trivial sense, and Ivanovsky is quoted as his competitor, as having been the first. Beijerinck himself was more gracious than later historiographers in acknowledging that he did not know about Ivanovsky's earlier publication, and he gave him credit. Ivanovsky, however, related that he had "succeeded in evoking the disease by inoculation of a bacterial culture, which strengthened my hope that the entire problem will be solved without such bold hypotheses" [4]. In 1903, when further criticising Beijerinck's conclusion about the *contagium vivum fluidum*, Ivanovsky claimed it to be a *contagium vivum fixum*. He wrote: "…the persistence of infectivity of the filtered sap can only be explained by the assumption that the microbe produces resting forms…" (spores). All these quotes demonstrate that Ivanovsky did not grasp the scope of his observations, that in his mind Koch's Postulates had fossilised into dogma [2].

When assessing achievements of the early workers, who we would call virologists today, one should avoid the trap of anachronism; it is a semantic trap. Thus, "virus" meant something quite different to Ivanovsky and Beijerinck, to Löffler and Frosch, to Reed and Carrol, than it means to us. "Fluid" at the turn of the century was synonymous with "non-corpuscular" insofar as particles with dimensions were concerned that could not be visualised by light microscopy – electron microscopy not having yet been invented. It took another forty years to demonstrate the particulate nature of virions.

The beginnings of animal virology in Germany

At the same time, filtration experiments were also performed with an animal pathogen in Germany, which lead to the identification in 1898 of the cause of foot and mouth disease (FMD) as a "filterable" or "invisible" virus. The finding resulted from a close collaboration between Friedrich Löffler, professor and director of the Institute of Hygiene in Greifswald, and Paul Frosch, then employed at Robert Koch's Institute of Infectious Diseases in Berlin; Löffler had been Koch's assistant there, until his appointment to the Greifswald chair in 1888. In 1890, Robert Koch already had deplored the fact that many infectious diseases were still aetiologically undefined; at the occasion of the 10th International Congress of Medicine in Berlin he proclaimed "…I tend to believe that the diseases mentioned (he referred to influenza, pertussis, trachoma, yellow fever, rinderpest, pleuropneumonia) are not caused by bacteria but by structured disease agents that belong to quite different groups of micro-organisms."

The optimistic atmosphere at the turn of the century, the enthusiasm about discovering more – perhaps even all – human and animal pathogens is reflected in the minutes of the 7th International Veterinary Congress in Baden-Baden, 7–12 August 1899. It was held under the protectorate of His Royal Highness the

Grand-Duke Frederick of Baden, and this is how Friedrich Löffler's report (in its original translation) reads for Tuesday, August 8th:

"The necessary funds were granted by the German Empire and the Prussian State, and I was charged with the execution of the work, which at first I carried on in the Institute for Infectious Diseases in Berlin, afterwards, in that of Hygiene at Greifswald, with the assistance of Professor Frosch, and later, from January 1898, of Dr Uhlenhuth.

When I undertook the work, the aetiology of foot and mouth disease was little studied. It was known that the disease was transmitted to cattle, pigs, sheep and goats, and that its germs might be carried by diseased animals and also by persons who had been in contact with them. The mode of action of the germ and the ways of infection were unknown.

The microscopical examination of coloured and not coloured preparations, the various methods of cultures did not permit us to discover the virus in the fluid, where it ought to have been found, namely, in the contents of the aphthae.

However, an entirely new and very interesting fact could be established. In order to see whether the contents of the aphthous vesicles, when filtered and attenuated with water, would grant immunity, they were passed through filters, which would with certainty hold back the most minute micro-organisms, for instance the bacilli of influenza. Still, the germ of aphthous fever did pass. In this way we were able to obtain a pure virus and to obviate any accidents that might arise from the presence of the organisms in the fluid that we used."

In view of the semantic trap mentioned above it should be noted that Löffler used the word 'virus' in the generic sense. Since antiquity the term has been applied to denote slime, animal semen, foul odour, acrid and salty taste, snake and scorpion venom, and poison in general; an early quote can be found in Cicero's *De amicitia* (On friendship, written about 45 B.C.) where a person's "...virus acerbitatis..." may be translated as "... the venom of bitterness" [3].

Thus animal virology originated at the same time as plant virology, and it took only four more years before the viral nature of yellow fever, an arthropod-borne infection, was determined. Animal virology arose from the need to control a disease of economic importance, as exemplified above, and Friedrich Loeffler was less concerned with the properties of the foot-and-mouth agent than with its elimination from the Prussian cattle population.

Importance and impact of veterinary virology

Before commenting on the importance and impact of "veterinary virology" in Germany, some definition is required. This is where ambiguity starts. Friedrich Loeffler had a medical education, as had Paul Frosch, though he held the chair for Hygiene at the Berlin Veterinary School during the last twenty years of his life. Is veterinary virology that branch of the discipline to which persons with a veterinary education have contributed? Then the fundamental studies at the Max-Planck Institute for Virus Research in Tübingen by Werner Schäfer – a vet by training – on murine retroviruses would fulfil the criterion. Or is veterinary virology aimed at companion and farm animals, as medical virology is aimed at human health? Then Erich Traub's studies at the Federal Research Institute for Animal Virus Diseases in Tübingen (FRIAVD) on murine lymphocytic choriomeningitis virus should be excluded...

This is a moot point, of course. To approach the topic in a formal way, I shall examine the chairs, directors and presidents of veterinary teaching and research establishments in Germany. However, as every historian will confirm, chronological distance is a prerequisite for a fair assessment of the past. Also, I call persons in this group my friends, and I would not want to sacrifice this relationship by giving too little credit to one, or – perhaps worse – too much to his adversary. I therefore decided to do a 30-years literature analysis starting in the 1960ies, when veterinary virology took off.

In doing this, I analysed the literary production of the past professors of virology at the veterinary faculties in Giessen, München and Hanover and the presidents of the FRIVAD in Tübingen, under the assumption that their leadership is reflected by co-authorships of articles. This should suffice to define the German aspect.

I should also specify the distinction I make between 'importance' and 'impact'. An important finding would be one that may be or has been useful for veterinary medicine. This is difficult to formally assess – perhaps a review of filed patents and their applications in products that have reached the market would be a method. A finding with a high bibliometric impact, on the other hand, has contributed to the science of virology in general, irrespective of its applicability – it suffices that it is interesting for virologists. This distinction is both arbitrary and fuzzy, but it does follow bibliometric terminology.

Veterinary virology units in Germany

The most venerable institution dedicated to the teaching of infectious and epidemic animal diseases can be found at Munich University. It had been part of General Veterinary Pathology since 1790, later baptised 'Institute for Microbiology and Infectious Diseases of Animals', and was headed by Anton Mayr from 1963 to 1990 (Fig. 1). Having been trained as a virologist at the Bavarian Vaccine Establishment and at the FRIAVD in Tübingen, which he led for 4 years, he maintained the general microbiology perspective during his entire career. Toni Mayr's continuous interest was in the poxviruses, in a broader sense in their role as inducers of non-specific immunologic defence. He coined the term 'Paramunität' – a peculiarly German invention – and developed parapox- and avipoxvirus preparations to stimulate the non-specific defence. Though poorly defined, products developed from these studies reached the marketplace and were much used as an anti-infectious panacea, again mainly in Germany and the Netherlands. Mayr's impact on the animal health scene in Germany has been remarkable and multifaceted. The students loved him, he was a sought after speaker at veterinary conventions, a relentless advocate of 'practical virology', a prolific writer of articles and handbooks, a politician – certainly the most general 'veterinary microbiologist' amongst the key figures here discussed.

The first dedicated chair of virology at a veterinary faculty was established in 1964 in Giessen, and Rudolf Rott (Fig. 2) become its head. During the six years preceding his appointment, he had worked with Werner Schäfer at the

Fig. 1. Anton ('Toni') Mayr   **Fig. 2.** Rudolf ('Rudi') Rott

Max-Planck-Institut für Virusforschung in Tübingen. The ortho- and paramyxo-viruses should stay with him during his entire scientific life, until (and after) his recent retirement. In his laboratory, he established groups working on alpha-, flavi- and birnaviruses, and on Borna virus – a German favourite, which only Hanover succeeded to ignore. The groups directed by Rudi Rott have made seminal contributions to general virology, and he had a clear conception about what scientific quality means.

Rudi Rott was much admired, revered – and feared. When he entered a discussion, everybody held his breath. He was a relentless critic, very influential in the German science environment – especially in the German Research Council – but he also determined the face of German virology internationally.

In keeping my promise to avoid comments on acting Heads of Departments, I skip the Veterinary Faculty at Berlin, where virology assumed independent status as late as 1978. Its first professor, Hanns Ludwig, is one of Rudi Rott's disciples, as is Hermann Müller, who recently occupied the chair in Leipzig. By establishing a school of virologists, Rudi Rott followed Werner Schäfer's example; Schäfer's disciples eventually occupied seven chairs of medical and veterinary virology in German-speaking Europe (Giessen, Hanover, Wurzburg, Cologne, Heidelberg and Zurich).

The Veterinary School in Hanover appointed Manfred Mussgay (Fig. 3) as its first full professor of virology in 1964. Having worked on Venezuelan equine encephalitis and vesicular stomatitis virus as a visiting scientist in Gernot Bergold's lab in Caracas, he further exploited the alphavirus model, gradually focusing on

Fig. 3. Manfred Mussgay   **Fig. 4.** Bernd Liess

the characterisation of pestiviruses. In 1967, he left for Tübingen to become President of the FRIAVD; the administrative duties there made it difficult to continue hands-on research, but he went through the showers almost every day and supervised work mainly on foot-and-mouth disease virus and murine and bovine retroviruses. Manfred Mussgay was a meticulous experimental worker, a cheerful personality with a contagious laugh. His death in 1982 at the age of 55 years was a severe loss not only for his friends, of which I have been one, but also for virology.

After Manfred Mussgay had left Hanover, Bernd Liess (Fig. 4) became his successor. He, too, had a spell in a tropical country, having worked in Kenya with Walter Plowright on rinderpest virus. Upon his return to Germany, he focused on the pestiviruses causing swine fever and bovine viral diarrhoea/mucosal disease. This line of research determined the profile of the Hanover laboratory until today. However, his interest in morbilliviruses continued, and he co-authored articles on canine and phocine distemper viruses. Bernd Liess retired in 1996 and was followed by Volker Moenning, a second-generation disciple of Schäfer's and former FRIAVD president.

Bibliometric analysis

If the importance of virology in Germany for the veterinary profession is difficult to assess, as mentioned, its impact can be estimated. To get an impression of the gross number of publications dedicated to veterinary medicine and virology, respectively, I queried the time-unlimited PubMed Entrez database (http://www.ncbi.nlm.nih.gov/htbin-post/Entrez/query?) of the U.S. National Library of Medicine/National Institutes of Health. MeSH (Medical Subject Headings) major topics and subheadings were used in the search, respectively. MeSH is a vocabulary of medical and scientific terms assigned to documents in PubMed by a team of experts. It is used for indexing articles, for cataloguing books and other holdings, and for searching MeSH-indexed databases, including MEDLINE.

From 1963 onward, about 137,000 "veterinary" articles were found as compared to about 208,000 papers containing "virus OR virology", both starting in 1963. The publication dynamics show a gradual increase in both categories in the late 1960's – most likely as a consequence of both the growth in funding of virus research and the increasing bibliometric activities of the Institute of Scientific Information (ISI). In the 1970–1983 period the ratio of virology/veterinary science papers remained fairly constant, with fewer indexed publications in the veterinary and medical sciences, with all their facets, than in virology alone. A conspicuous divergence occurred afterwards. In 1984, the retrovirus that causes AIDS was discovered, testing for antibodies was begun, HIV research took off, and many virologists jumped on the bandwagon that was propelled by a superabundance of funding.

The fraction of veterinary papers in the virology category was assessed by querying "((virus[MeSH Major Topic] OR virology [MeSH Major Topic]) AND veterinary [Subheading])", which resulted in 9707 hits – in other words: 4.7% of all indexed publications in virology contain the term "veterinary". When extending the query with "Germany" the number is reduced to 91. These figures are nothing but indicators, as may be expected from such a superficial analysis; thus "veterinary" may be absent from many papers on viruses affecting animals (resulting in a underestimate).

Subsequently, the names of the German virologists mentioned above were used to search PubMed; a steady stream of 22 ± 5 papers/year from 1968 to 1995 shows the productivity of the groups. The articles have appeared in the journals in Table 1, with additional bibliometric indicators. 'Total cites' is the number of times that each journal has been cited in a given year. The impact factor (IF) is a measure of the frequency with which the average article in a journal has been cited in a particular year (the number of current citations to articles published in a specific journal in a two year period divided by the total number of articles published in the same journal in the corresponding two year period). The immediacy index (II) is a measure of how quickly the average article in a specific journal is cited.

The following considerations are meant to provide "food for thought". Of the 640 publications examined, about 3/4 have appeared in 12 journals (in total, 91 journals have been used for publication). Only "virology" journals rank amongst the first 12, while the "veterinary" category journals are generally low ranking. A notable exception in Veterinary Microbiology, which (in the 1995 listing) ranks 7th in the "veterinary" subject listing, though only 38th in the "microbiology" category. The dichotomy between excellent virology journals and low-impact

Table 1. Listing of publications from the authors Liess, Mayr, Mussgay, Rott and Traub in the 1968–1995 period, with the numbers(#) of papers, total cites, journal impact factors (IF) and immediacy indices (II)

Total Publications with >1 pub./journal by Liess, Mayr, Mussgay, Rott, Traub, as listed in MEDLINE
(Ordered according to frequency)

	Journal	#	total cites	IF	II	Vir	Vet	Mic	Bio/MDS
1	Zentralbl Veterinarmed (J Vet Med B)	122	394	0.460	0.079	–	38	–	–
2	Arch Gesamte Virusforsch & Arch Virol	58	2768	1.384	0.323	12	–	–	–
3	Virology	54	23475	3.901	0.674	3	–	–	–
4	Deutsche Tierarztl Wochenschr	44	445	0.231	0.009	–	64	–	–
5	J Gen Virol	41	12589	3.410	0.444	5	–	–	–
6	Berl Munch Tierarztl Wochenschr	35	294	0.234	0.026	–	63	–	–
7	Zentralbl Bakteriol	27	1018	0.898	0.014	–	–	44	–
8	Tierarztl Prax	24	–	–	–	–	?	–	–
9	J Virol	20	45077	6.033	1.176	1	–	–	–
10	Med Microbiol Immunol (Berl)	20	587	2.145	0.136	–	–	19	–
11	Vet Microbiol	15	1516	1.076	0.202	–	7	38	–
12	Bull Off Int Epizoot	11	–	–	–	–	–	–	–
13	Fortschr Med	7	–	–	–	–	–	–	?
14	Intervirology	7	823	1.260	0.037	13	–	–	–
15	EMBO J	6	59817	13.505	2.281	–	–	–	5
16	Nature	6	257287	27.074	6.043	–	–	–	1

The right 4 columns show the ranking of the respective journals in in the indicated bibliometric categories (J Virol ranks 1st in the 'virology' category, Vet Microbiol ranks 7th in the 'veterinary' category etc.)

veterinary journals becomes even more pronounced, when "the German specialities" are compared with the rank listing in virology and the veterinary sciences (in the latter category only journals that would publish microbiological papers have been listed; Table 2). The language bias may have contributed to the skewed distribution that makes German journals rank behind for example Scandinavian, Czech and Belgian ones.

Veterinary virology publications from Germany in this time frame reveal the fascination by virion structure, physical characteristics and structure-function relationships – arguably Werner Schäfer's heritage. Amongst journals used only once by an author are very prestigious ones such as Cell (Rott), and many titles that are marginal to virology. Though "veterinary" would suggest interest in the animal's role in viral infections, there is little work published in journals dedicated to immunology and pathogenesis, e.g. no papers in Vet Immunol Immunopathol; Am J Vet Res, or Proc Soc Natl Acad Sci USA.

A comparison of immediacy indices reveals the dynamics in the various bibliometrical fields: if it took an article published in the Zentralblatt für Veterinärmedizin one year to be quoted, then an author publishing in the *Journal of Virology* would be cited within a month – an author writing in Cell 4 days. This is an arithmetic exercise, of course, but quite illustrative.

Table 2. Comparative subject category listing of journals where virologists might want to publish, with their rankings in the bibliometric categories 'virology', 'veterinary sciences', and 'molecular biology/immunology'

Comparative Subject Category Listing (SCI Journal Citation Reports 1995)

Virology			Veterinary Sciences			Mol. Biology/Immunology	
Rank	Title	IF	Rank	Title	IF	Title	IF
1	J Virol	6.033	5	Vet Immunol Immpath	1.138	Cell	40.481
2	Adv Virus Res	5.120	7	Vet Microbiol	1.076	Nature	27.047
3	Virology	3.901	11	Vet Rec	1.014	Immunol Today	25.228
4	Semin Virol	3.625	15	Am J Vet Res	0.907	Science	21.911
5	J Gen Virol	3.410	16	Vet Pathol	0.879	J Expl Med	15.126
6	J Med Virol	2.232	19	Avian Dis	0.774	EMBO J	13.505
7	Virus Res	2.161	24	Res Vet Sci	0.717	PNAS	10.520
8	Antivir Res	1.849	25	J Comp Pathol	0.715	J Immunol	7.412
9	Rev Med Virol	1.780	28	Comp Immunol Microb	0.645	J Biol Chem	7.385
10	Virus Genes	1.472	30	Aust Vet J	0.627		
11	J Virol Methods	1.464	36	Adv Vet Sci Comp Med	0.516		
12	Arch Virol	1.384	38	J Vet Med B	0.460		
			The German Specialties...(<0.250)				
			62	Wiener Tierarztl Monat			
			63	BMTW			
			64	DTW			
			68	Schweiz Arch Tierheilk			
			70	Tierarztl Umschau			
			76	Monatsh Veterinarmed			
			87	Prakt Tierarzt			
			89	Kleintierpraxis			

Another priority issue

While Robert Koch and Louis Pasteur have become household names, so to speak, in microbiology, another figure in the virology, immunology, vaccinology triangle has been almost completely forgotten. I should like to draw the veterinary virologists' attention to a self-taught Dutchman, a miller and farmer, who is still remembered in his birthplace. A monument was recently erected in Winsum/Friesland to honour Geert Reinders (1737–1815), the 'inoculator' and saviour of the country from rinderpest. After the 1768 epidemic in the Netherlands he concluded

- that cattle which had experienced the natural illness were protected from disease after another infection
- that the same was true for animals with only light symptoms e.g. after vaccination, and

- that the mode of inoculation and supportive therapy had no influence on the outcome of infection. He also discovered what we today would call "maternal immunity", the protection transferred from an immune cow to its calf.

Geert Reinders published his observations in 1776 – Edward Jenner's vaccinia protection experiments appeared in press two decades later. At that time, however, Jenner was already a public figure, known as a skilful and popular surgeon, eventually becoming a member of the Royal Society due to his discovery of the nesting parasitism of the European cuckoo. Reinders' findings were published in Dutch and had a small readership. Historically, it would appear that veterinary vaccinology predated medical vaccinology – as veterinary virology preceded medical virology. The speed of progress, however, was quite different.

Scientific priority is of historiographic interest (where chauvinist motives may obfuscate the issue), but above all it is of importance to every scientist. However, being first chronologically is different from the priority perceived by the scientific incrowd, by academia, by the public. It takes social and political skills to convince the 'shakers and movers', the establishment, the referees of high-ranking journals, that one really has made a novel finding. Proverbially, only posterity will assess and acknowledge the inventor and the discoverer. The book on the rediscovery of viruses will eventually be written, probably by a retired professor, who saw his favourite finding go unnoticed, only to return in another countenance, published and publicised by a dynamic young colleague from a renowned research establishment.

There is no doubt that the cradle of virology was rocked about 100 years ago, in Russia, Prussia, and in the Netherlands; nor is there any doubt that a veterinary problem led to one of the greatest serendipitous discoveries in biology.

References

1. Beijerinck MW (1899) Über ein Contagium vivum fluidum als Ursache der Flekkenkrankheit der Tabaksblätter. Zentralbl Bakt Abt II 5: 27–33
2. Bos L (1981) Hundred years of Koch's postulates and the history of etiology in plant virus research. Neth J Plant Pathol 87: 91–110
3. Klotz R (1857) Handwörterbuch der lateinischen Sprache. G. Westermann, Braunschweig
4. Ivanovsky D (1899) Über die Mosaikkrankheit der Tabakspflanze. Centralbl Bakteriol II 5: 250–254
5. Ivanovsky D (1903) Über die Mosaikkrankheit der Tabakspflanze. Z Pfl Krankh 13: 1–41
6. Mayer A (1886) Ueber die Mosaikkrankheit des Tabaks. Landw VersStn 32: 451–467 [English translation (1942) Concerning the mosaic disease of tobacco. Phytopathol Classics 7: 11–24]

Authors' address: Dr. M. C. Horzinek, Veterinary Faculty, Utrecht University, Virology Unit, Yalelaan 1, 3508 TD Utrecht, The Netherlands.

The evolution of viruses, the emergence of viral diseases: a synthesis that Martinus Beijerinck might enjoy

F. A. Murphy

School of Veterinary Medicine, University of California, Davis, California, U.S.A.

Nothing in life makes sense,
except in the light of evolution.
T. Dobzhansky, 1973

Summary. The relentless production of viral variants and their selection for improved "fit" are seen from the perspective of the infectious disease sciences as ever-changing viral phenotypes and emerging disease risks. In the Darwinian cause:effect equation, we can characterize very well the effects of mutation and selection – these are catalogued as new viral phenotypes or pathotypes. However, the selective forces themselves driving such changes remain rather mysterious. Many selective forces must be at work, acting on the virus, the host, the host population and the environment. In some instances the virus seems to test new unoccupied niches in the absence of any apparent environmental change, but usually it is clear that changes are driven by human activity. Most important must be the ever increasing density of human, domestic animal and crop plant populations and the consequent increased opportunities for transmission of viral variants. Also important must be the great changes affecting all ecosystems – these especially favor the emergence of new zoonotic viruses and viral "species jumpers." The great increase in human travel and transport carries exotic viruses, vectors and hosts around the world, again favoring viral occupation of new niches. The rise of bioterrorism adds yet another threat. Increasing numbers of emerging viral disease episodes seem to be linked to a decline in global resources for proven public health programs, agricultural extension programs, and the like, programs that have stood in the way of the spread and evolution of viral pathogens. If the relationship between viral evolution and the emergence of new viral diseases is rooted firstly in the host and the host population, then more research and resources must be directed to intervention at these levels rather than at the level of the viruses themselves.

Introduction

Our understanding of the direction and rate of the evolution of viruses follows upon the most spectacular achievements in viral genetics and viral genomics. Of course, our understanding is at its best when we consider the smallest variations between viruses, say variations between strains or species in the same genus. It is quite a different matter to discuss viral evolution in the context of the evolving natural history of pathogens, where important phenotypic characters, dominant selective forces and the true essence of "fit" (i.e., reproductive success) are much more complex and still quite mysterious. At this level there seems to be a conflagration of Darwinian determinants driving viral evolution – virologic determinants (such as mutation, recombination, reassortment, natural selection, fitness adaptation, evolutionary progression), host physiologic determinants (such as host innate, immune and inflammatory responses), natural environmental determinants (such as ecologic and climatic determinants), and determinants pertaining to human activity (such as behavioral, societal, commercial, transport, iatrogenic, and malicious determinants). Would not Martinus Willem Beijerinck, fresh from 100 years of observing the march of virologic research, enjoy being here today as this is discussed?

Focus upon the viruses themselves leads to a myopic view of important larger subjects, subjects such as viral natural history and the emergence of new viral diseases. Although in some instances the emergence of new diseases has seemed to follow solely upon mutation in the virus, in most instances determinants external to the virus seem to be paramount. How can we fathom the relationship between viral evolution and disease emergence? The answer may still ultimately lie in the field of molecular genetics, but this must include host genetics, host immunogenetics, and host population genetics, as well as viral genetics.

Insight into the interrelationship between virus evolution and the emergence of new viral diseases must stem from observations of nature – it must stem from study of exemplary viral diseases in their natural settings. When we conduct such studies, we see in some cases endemic constancy, in other cases a waxing and waning in disease incidence, in yet other cases spectacular de novo epidemic explosions, and in every case a sense that the next "new" virus and the next emergent disease episode will be as unpredicted as the last. Such studies, the subject of this paper, follow the tack advocated by Stephen Jay Gould: "The beauty of nature lies in detail; the message in generality. Optimal appreciation demands both, and I know no better tactic than the illustration of exciting principles by well-chosen particulars."

Darwinian forces at work in the evolution of the virus

All virologists appreciate full well that viruses undergo an infinitely long series of replication cycles as they are transmitted from host to host, and that during this process spontaneous mutants are continually generated, some of which produce variant phenotypes. In the Darwinian sense, however, continuing, relentless production of mutants is just the fodder for the selection of the occasional viral variant with improved "fit." The great mystery concerns the selective forces themselves – in the Darwinian cause:effect equation, we understand the second term rather well, but not the first.

Selection of particular viral variants must take place chiefly in vivo, within infected cells of the host, during replication. Selection must predominantly and preferentially affect the ability of the variant to replicate and be serially transmitted, but in some cases it must also affect virion qualities pertaining to tissue tropism, host range, and environmental stability, as these affect perpetuation of the virus in nature. There are few data to support more specificity in this matter. The great evolutionary scholars of the day rarely mention organisms that reproduce asexually, and in most cases never mention the viruses at all – so, it seems that we are on our own. We each have our own perspectives, our own biases, to guide us in considering the selective forces that have guided the evolution of our favorite viruses; further, our view of these forces is biased by our sense of those viral characters that are most important in the "success" of our favorite viruses. Whatever our perspective, perhaps the watchword is, "never underestimate the power of selection."

In considering the effects of selective forces on virus evolution, it would seem that some of the most successful viruses have evolved.

The capacity to replicate rapidly

In many instances, the most virulent strains of a virus replicate faster than more temperate strains (e.g., enteropathic strains of mouse hepatitis virus replicate more rapidly than more temperate strains). This character seems to define the most successful variants of even those viruses with the slowest cycles in nature (e.g., human immunodeficiency virus 1). However, if replication is too rapid, it can be self-defeating – extremely rapid viral growth may not allow time enough for transmission before the host is removed by death or severe illness (e.g., myxoma virus in rabbits in Australia, where strains with intermediate replication qualities were selected within a few years of initial release of a very virulent strain).

The capacity to replicate to high titer

Very high vertebrate host viremia titer is employed as a survival mechanism by arthropod-borne viruses to favor infection of the next blood-feeding arthropod (e.g., epidemic strains of Venezuelan equine encephalitis virus produce very high viremia titers in horses). The same viruses produce very high titers in the salivary glands of their arthropod hosts so as to favor infection of the next vertebrate host. Such very high virus titers can be associated with silent infections in some natural vertebrate hosts (e.g., eastern and western equine encephalitis viruses in reservoir avian hosts), but in those vertebrate hosts that we care most about the evolution of this capacity is often associated with severe, even fatal, illness (e.g., eastern and western equine encephalitis viruses in humans and horses).

The capacity to be shed quickly

Not all viruses that replicate rapidly are shed rapidly, but those that are all seem to employ the simplest viral "entry/infection/exit cycle," where all aspects of infection take place in the same superficial target cells (e.g., parainfluenza viruses, which nearly exclusively infect airway epithelial cells). This infection pattern often does not stimulate a vigorous host response and when it does it only appears after transmission has already been accomplished (e.g., rotaviruses in the intestinal epithelium).

The capacity to be shed for long periods of time

The evolution of the capacity for chronic shedding offers exceptional opportunity for virus survival and entrenchment (e.g., maedi/visna virus in sheep, in which persistence is so sustained that in Iceland eradication has required synchronous depopulation of whole districts). Recrudescence and intermittent shedding add additional survival advantages to some viruses (e.g., varicella-zoster virus); long-term shedding from congenitally infected hosts represents yet another survival advantage (e.g., rubella virus).

The capacity to restrict gene expression

Viral latency may be maintained by restricted expression of genes that have the capacity to kill the cell. During latent infection, some viruses express only a few early genes that are necessary in the maintenance of latency (e.g., herpesviruses); during reactivation the whole viral genome is transcribed again. This strategy protects the virus from all host defenses except during recrudescence.

The capacity to cause non-cytocidal infection

Some viruses establish chronic infections without killing the cells in which they replicate (e.g., arenaviruses and hantaviruses in their reservoir rodent hosts, retroviruses in virtually all hosts). The capacity to infect resting cells represents an extension of this survival advantage (e.g., Epstein-Barr virus in B-lymphocytes). Similarly, the capacity to infect undifferentiated cells presents yet another extension of this strategy (e.g., papillomaviruses, which invade basal cells of stratified epithelium but produce infectious virions only in fully differentiated cells near the body surface).

The capacity to replicate in certain key tissues

The evolution of viral tropisms and the employment of specific host cell receptors are major determinants in defining viral disease and transmission patterns (e.g., rabies virus, which employs the acetylcholine receptor at neuromuscular end organs). Infection in sequestered sites provides great survival advantage: such sites include the central nervous system (e.g., rabies virus in neurons, as the cause of fury), the kidneys (e.g., Sin Nombre virus in *Peromyscus maniculatus*, its reservoir host), the salivary glands (e.g., Machupo virus in *Calomys callosus*, its reservoir host), the lymphoid system (e.g., human immunodeficiency viruses 1 and 2), and the reticuloendothelial system (e.g., measles virus).

The capacity to elude the host immune response

Vertebrate hosts have evolved elaborate immune systems to defend themselves against the viruses, but viruses have in turn evolved systems to elude host defenses. Viruses, particularly those with large genomes, encode proteins that interfere with specific host anti-viral activities (e.g., adenoviruses encode a protein that binds to MHC class I protein, reducing its cell surface expression). Some viruses cause syncytia, enabling the viral genome to spread from cell-to-cell without exposure to host defenses (e.g., mumps virus in salivary glands). The capacity to cause immunologically tolerant infection represents an evolutionary progression that gives some viruses an extreme survival advantage (e.g., bovine viral diarrhea virus).

The capacity to evoke an immune decoy and mask viral epitopes

Some viruses have evolved strategies for evading neutralization by the antibody they elicit (e.g., Ebola virus produces a truncated version of its peplomer glycoprotein which is secreted extracellularly and "soaks up" antibody). The glycoproteins of some viruses are very heavily glycosylated – carbohydrate may constitute one-third of the mass of surface peplomers – thereby masking epitopes on virions and virion budding sites on infected cells (e.g., Ebola virus, Lassa virus, Rift Valley fever virus).

The capacity to evade host herd immunity by genetic/antigenic drift and shift

Mutations (point mutations leading to drift, recombination or reassortment leading to shift) may be the cause of viral escape from a level of immunity in a host population that would otherwise interrupt transmission. Of course, influenza is the example par excellence – the survival advantage of these capacities for the influenza viruses is evident in the history of pandemics in humans and epidemics in chickens in high-tech poultry industries. Similarly, seemingly minor point mutations have been the basis for viral species jumping (host range extension) (e.g., canine parvovirus 2 emerged by mutation of from feline panleukopenia virus).

The capacity to survive by killing (or conversely, not killing) the host

Whether or not a virus regularly kills its host must reflect a central survival strategy, but the lesson so often taught to students, that evolutionary progression always favors viral commensalism, seems simplistic. Again, the emergence of myxoma virus variants of intermediate virulence in Australia is often used in this lesson, but the term "intermediate virulence," as used here, is relative – the long-term surviving variant virus in Australia still kills about 50% of exposed rabbits (vs. 90–99% in the year after virus was first released). Rabies virus evolution seems to favor lethality in reservoir hosts – such hosts increase breeding rates and usually refill niches quickly with susceptible young hosts – this favors the perpetuation of the virus. Yellow fever virus has seemingly remained constant in its capacity to kill its human hosts over 300 years. Human immunodeficiency virus 1 is also extremely virulent, killing a very high proportion of infected, untreated subjects, again without any evidence of becoming more temperate. Viruses that kill their hosts after assuring adequate or even maximal transmission do not represent failures in "fitness."

The capacity to be vertically transmitted by integration of the viral genome into the host cell genome

Viruses that employ vertical transmission via the integration of proviral DNA into the genome of host germ-line cells are perpetuated without ever confronting the external environment (e.g., endogenous retroviruses). This represents another evolutionary progression, fortunately one not associated with any important human pathogen.

The capacity to survive after being shed into the external environment

All things being equal, a virus that has evolved a capsid that is environmentally stable must have a substantial evolutionary advantage (e.g., canine parvovirus 2, which was transported around the world within two years of its emergence, mostly by fomite carriage).

This list of capacities of various successful viruses may seem overly long and convoluted, but in fact more items could be easily added. Moreover, many successful viruses employ several of these capacities, each acting synergistically to favor transmission between hosts and perpetuation in host populations. Worse yet, many capacities that we think of only in terms of viral transmission and perpetuation correspond to capacities associated with virulence. In some cases, this is just a matter of whether one is thinking as an epidemiologist or as a pathologist. What do these capacities of successful viruses suggest about specific selective forces at work in nature? What do these effects suggest about their causes? Does the diversity of these capacities indicate an equally diverse set of forces contributing to selection for "fitness"? Most importantly, *what are the selective forces that were involved in the evolution of those viruses that represent the most significant pathogens of today, and what are the selective forces that will be involved in the evolution of the emergent pathogens of tomorrow?*

Selective forces operating in nature, whatever their nature, seem to be attuned to the level and rate of change that can be tolerated by viral genomes – excesses are self-destructive – the *status quo* is overwhelmed – failures disappear from the gene pool. "Fitness" for survival in nature represents the fine balance of many traits. Interestingly, most experimental manipulations of viruses aimed at testing hypotheses pertinent to this subject cause a loss of "fitness." Laboratory-passaged strains of viruses are often faint shadows of their wild type progenitors. There are many experimental approaches being taken today that require that artificially constructed variant viruses approximate the wild type in their "fitness." New vaccine strategies, eukaryotic gene vectoring, even the notion of replacing in a particular econiche a pathogenic virus by a non-pathogenic variant, require not only the genotypic stabilization that can be achieved with infectious molecular clones, but also an understanding of the selective forces that can stabilize or destabilize the genotype. Perhaps present experimentation will have a spin-off; perhaps it will provide guidance as we try to unravel the mysteries of selection.

The quasispecies concept, advanced by Manfred Eigen, John Holland and their colleagues, has greatly advanced our notion of how selective forces operating at the level of the virus population may influence the rate and direction of evolution. In the quasispecies concept, the virus species, defined by conventional phenotypic properties, exists as a genetically diverse, dynamic, competing population of variants, each having only a fleeting existence. Taken together, all the variants resemble a metaphorical cloud – the quasispecies cloud – where variants probe the limits of their environment (i.e., their "sequence space"). Over relatively short time periods genotypic drift occurs as particular variants gain advantage; over longer time periods drift leads to the evolution of substantially different viruses, that is new strains and species. For example, the quasispecies cloud may yield immune escape variants or species jumpers (e.g., human immuno-deficiency viruses 1 and 2 are clearly the products of species jumping). Viral evolutionary progression is also affected greatly by other population-based phenomena, such as genetic bottlenecks, Muller's ratchet (mutations cumulatively and irreversibly eroding fitness in ratcheted fashion), random drift, and perhaps even punctuated equilibrium.

It had long been held that selection favors or discriminates against phenotypes, not genes or genotypes, but this point has been argued hotly in recent years. In sexually reproducing organisms, some evolutionary biologists have contended that the unit of selection is the gene (e.g., in the concept of "the selfish gene"), while others have maintained that it is the intact, reproducing organism (e.g., in the concept of "genetic altruism"). At the same time, most of these authorities have agreed that since in asexually reproducing species the parental genome is reproduced in all progeny, the genotype and phenotype are co-variant, and the unit of selection must be the organism (or virus) itself. Now, consequent to the rise of the concept of quasispecies this notion must be refined: the unit of selection is not the most fit genotype (i.e., the master sequence); rather, it is the entire quasispecies cloud of variants that is acted upon by selective forces and yields the next master sequence, the new phenotype.

Darwinian forces at work in the reactive evolution of the host

One pervasive metaphor of evolutionary biology is that selection is like the process of fitting a key to a lock. The lock, in the case at hand the host and environment in which the virus must be perpetuated, is a fixed entity, and the key, in this case the virus, must be adjusted to fit – that is the niche provides the selective force or pressure and virus must adapt to the niche. This metaphor may be useful; however, in virology we know that the host as well as the virus evolves reactively. For example, many human populations evolved a substantial level of resistance to smallpox virus infection.

As we examine the progression of particular virus:host relationships, the metaphors of war and battle come to mind. At first glance, the armamentarium of the host, the accumulation of all the survival mechanisms inherited from its progenitors, seems pervasive – after all, the host brings a much more complex genome and many more gene products to the battle. On the other hand, if the Lilliputian genomes of the viruses were our only indication of the quality of their weaponry, our interests and concerns would long ago have turned to other, more significant threats to our survival and that of the animals, plants, and invertebrates upon which we depend. In fact, the viruses bring formidable weapons to the battle. The combatants, their weapons and their battle tactics are the stuff of medical and veterinary virology, pathology and immunology, and of plant pathology, invertebrate pathology and related sciences – the mass of the literature in these fields suggests great complexity.

A view of the weaponry of the viruses is described above. The weaponry of the host is usually categorized as: (1) innate, nonspecific, resistance factors (e.g., interferons, inflammatory cytokines); (2) acquired, specific, resistance factors (e.g., the cellular and humoral immune systems); (3) physiologic factors affecting resistance (e.g., age, nutritional status, and hormonal status, especially in pregnancy); and (4) medical care factors (e.g., antiviral chemotherapy, immunoprophylaxis, and immunotherapy). The more innate, generalized weapons of the host may be equal in power to the more specific, acquired weapons – after all, they have been evolving for far longer – but, some are still rather mysterious and some are subject to very little ongoing research. Terms used by evolutionary biologists to describe the presence of such weapons, such reactive traits, in the host population include stability, persistence, longevity, fecundity, and fidelity. Although these terms are not widely used in virology or the infectious disease sciences, their inference is clear – evolutionary progression must include penetrance of reactive traits into the population at risk.

One vertebrate host weapon system stands out, that is the immune system. The capacity of the immune system reflects an incredible evolutionary progression, perhaps surpassed only by the evolution of the central nervous system. Its evolutionary progression has been driven by diverse threats to the survival of individual hosts and the host lineage – viruses, microorganisms, parasites, toxic chemicals, radiation, cancer cells, perhaps any foreign entity smaller than the host itself. Because the immune system evolved to deal with such diverse threats, we recognize that it cannot be perfect – indeed, it is Darwinian in its purpose and its capacities.

It has been said that some viruses have taken advantage of what have been called "weak links" in host immune defenses – this notion must seem even more anthropocentric than others mentioned in this paper! For example, we envision human immunodeficiency virus 1 having an uncanny intelligence as it attacks

CD4+ T lymphocytes, the central cells of the immune system. Similarly, we envision hepatitis B virus having devious qualities as it causes the persistent infection and immune tolerance that are the bases of chronic progressive cirrhosis and hepatocellular carcinoma. In contrast, is it not intriguing that we do not call a virus incompetent when it cannot stand up to host antibody, T cells or macrophages? Pursuing this notion further, it seems that rather than focusing on immunological "weak links," we might better focus on the power, the successfulness of the immune system. Nothing brings this point home better than examining the consequences of the immunosuppression caused by human immunodeficiency virus 1 or immunosuppressive anti-cancer drugs or congenital immunodeficiency diseases. The unleashing of opportunistic infectious agents, including viruses, as the immune system fails, leads to the most catastrophic clinical syndromes. Ordinary viruses become lethal. The pathologist, reviewing the course of cytomegalovirus infection in tissues from a fatal case of AIDS or varicella-zoster virus infection in tissues from a fatal case of pneumonia in a child with leukemia, is easily returned to a point of wonder in regard to the power of the immune system.

This sense of wonder is reinforced by review of studies with inbred mice, in which very large repertoires of genes that confer survival advantage upon the host have been identified. Of course, many of these genes map to major histocompatibility and Ir loci and, therefore, influence host immune responses to multiple viral and microbial infections, but others are specific for a single family of viruses and their functions are quite mysterious. Conversely, line-breeding and in-breeding have yielded classical strains of mice that are exquisitely sensitive to certain viruses – these are the strains that have been used for many years to isolate arboviruses, picornaviruses and rabies virus. While the nature of most resistance alleles in these mice is unknown, it would appear that their analogues in nature represent specific survival mechanisms that have been subject to natural selection over evolutionary time. It remains now to identify them, genotypically and phenotypically.

The subject of the evolution of host species that can survive the onslaught of pathogenic viruses seems too large, too enigmatic, to pursue further here. Understanding fails at the same point as when trying to understand virus evolution: what are the forces that were involved in the evolutionary progression that led to the most successful vertebrate species of today? What are the forces that will be involved in the selection of host species that can deal with the emergent pathogens of tomorrow? Can we determine how virus infections may have driven the evolution of the immune system? Does the complexity of the immune system, with its incredible ability to discern a seemingly infinite number of epitopic specificities, suggest anything about the limits of diversity of pathogenic viruses and microorganisms?

Darwinian forces involving the host population

The notion of selection occurring at the level of the group or population has acquired a sophisticated theoretical, conceptual base in recent years, but seemingly this base has been built upon only a few examples (notably, the barbed stinger of honeybees). However, if the concept is relaxed to include scondary influences upon evolutionary progression, one might conclude that selective forces that act primarily at the level of the host population might be more important than any others.

The host population is where the cumulative influence of basic human behaviors comes to bear, whether this is the personal behavior of humans within communities, or various aspects of societal behavior. For example, changing personal behavior affected by the peer community (e.g., multiple sex partners, intravenous drug usage) has led to increased transmission of sexually-transmitted and blood-borne virus diseases, and the inclusion of offal in feed supplements has been identified as the cause of the epidemic of bovine spongiform encephalopathy in cattle in the United Kingdom.

The host population is also where most ecological influences on the evolutionary progression of viruses seem to operate – nearly all ecological changes that have favored virus spread have been caused by human actions. In this regard, the increasing impact of the arboviruses following upon various ecological changes is exemplary:

- Population movements and the intrusion of humans and domestic animals into ancient arthropod habitats have resulted in dramatic epidemics. Some are of historic significance: the Louisiana Purchase came about partly because of losses Napoleon's army suffered from yellow fever in the Caribbean-several decades later the same disease halted the building of the Panama Canal.
- Ecologic factors pertaining to changes in unique environments have contributed to many new, emergent disease episodes. Remote econiches, such as islands, harboring distinctive species of potential hosts and vectors, are often particularly vulnerable to an introduced virus. For example, the initial Pacific "island-hopping" of Ross River virus in the 1980s from its original niche in Australia caused virgin-soil epidemics of arthritis-myalgia syndrome in Fiji and Samoa.
- Deforestation has been the key to the exposure of farmers and domestic animals to new arthropods and their viruses. The occurrence in recent years of Mayaro virus disease in Brazilian woodcutters as they cleared the Amazonian forest, is a case in point.
- Increased long-distance travel facilitates the carriage of exotic arthropod vectors and their viruses around the world. The carriage of the eggs of the Asian mosquito, *Aedes albopictus*, to the United States in used tires represents an unsolved problem of this kind – this mosquito is a proven vector foe dengue and other viruses.
- Increased long-distance livestock transportation facilitates the carriage of arthropods (especially ticks) and their viruses around the world. The introduction of African swine fever virus from Africa into Portugal (1957), Spain (1960) and Central and South America (1960s and 1970s) is a case in point.
- Ecologic factors pertaining to water usage, especially irrigation, are becoming important factors in virus disease emergence. The problem with primitive irri-

gation systems, which are developed without attention to mosquito control, is exemplified in the emergence of Japanese encephalitis in more and more areas of Southeast Asia. New routings of long-distance bird migrations, brought about by new man-made water impoundments, represent an important yet still untested additional risk for the introduction of arboviruses into new areas.

- Ecologic factors pertaining to uncontrolled urbanization are contributing to many new, emergent arbovirus disease episodes. Arthropod vectors breeding in accumulations of water (tin cans, old tires, etc.) and sewage-laden water is a worldwide problem. Environmental chemical toxicants (herbicides, pesticides, and residues) can also affect vector-virus relationships, directly or indirectly. Mosquito resistance to insecticides is a direct consequence of unsound mosquito abatement programs and insecticide usage against crop pests.

- Global warming, affecting sea level, estuarine wetlands, fresh water swamps, and human habitation patterns may be affecting vector-virus relationships throughout the tropics – however, data are scarce and many programs to study the effect of global warming have not included the participation of arbovirologists.

Qualities of human host populations, per se, that may affect the success of viral transmission and the perpetuation of viruses in nature include: (1) population size and density; (2) population age distribution; (3) population economic and nutritional status; (4) population educational status; (5) population vaccination and immune status (herd immunity); and the like. Host population qualities that may affect the success of animal, plant and invertebrate viruses seem analogous.

Particular viruses have evolved survival strategies to deal with the extremes in host population qualities – of the viruses transmitted from human to human, most thrive when introduced into a new human population, but many express additional demands. For example, some of the viruses that are maintained by aerosol, respiratory droplet or fecal-oral transmission depend on a minimum density of susceptible hosts to sustain their transmission chain. If the density of susceptible hosts falls below a critical threshold, the chain may be broken. This is exemplified by the spontaneous disappearance of measles from human populations less than 300,000 in size. This phenomenon led to the speculation that human measles virus must have emerged from some ancestral animal morbillivirus, such as rinderpest virus, only after the rise of civilizations and cities. Our sense of the importance of population density is bolstered by the success of vaccination campaigns, such as the global polio vaccination campaign. In such campaigns, transmission chains are often broken even when the level of herd immunity achieved is less than desired.

Of all population-based characters, population age distribution is recognized as one of the most important. For example, in dense urban populations in developing countries the transmission of many viruses occurs at a very early age and spread through the population is very rapid. This may be associated with a low disease:infection ratio, as with polio, or a high disease:infection ratio, as with rotavirus diarrhea and measles, but in every such instance the perpetuation of the virus is enhanced when young hosts are involved. As the average age at the time of infection increases, due to improved public health and/or community hygiene measures, disease incidence may become more or less common or severe, but again in every instance the success of the virus is threatened when it must employ older hosts than was the case historically. In turn, however, as such population-based qualities continue to exert their influence over time, it seems inevitable that they should drive the evolution of the viruses and eventually lead to the emergence of more '"fit" variants. The rise of measles in college-age populations, although not dependent upon any detectable viral variance, in a case in point. The fictional genetically-engineered, aerosol-transmitted Ebola virus of Hollywood fame touches the public's imagination in this regard.

There are, of course, many more facets to the evolution of host populations that relate to the evolution of viral pathogens, but again the subject becomes unsatisfying because we do not understand the nature of the selective forces at work here. Again, in the cause:effect equation we understand, at least a bit, the second term, the effect of evolution on the virus and host phenotypes, but not the first, the selective forces themselves.

Synthesis: The relationship between the evolution of viruses and the emergence of viral diseases

To return to the central questions posed in this paper: How can we fathom the relationship between the evolution of viruses and the emergence of viral diseases? What are the selective forces that will be involved in the evolution of the emergent pathogens of tomorrow? As one tries to merge ideas about the evolution of the viruses and experiences with recent emergent disease episodes, several predicate thoughts come to mind, all calling for better integration of various disparate "databases." First, we must integrate information on just how the viruses are changing in nature – change, evolutionary progression, is in the nature of the beast. Second, we must integrate information coming from viral genomic sequencing (and partial sequencing) and we must move from viral genomics to functional genomics. Third, we must integrate information from representative animal model studies – this is an essential intermediate stop between basic and clinical sciences, the place occupied by the fields of viral pathogenesis and pathophysiology. Fourth, we must integrate information coming from studies of host populations – this is another key intermediate stop, the place occupied by the fields of epidemiology and clinical medicine – most importantly, here is where we may consider new approaches for intervening in the course of emerging diseases. Finally, we must integrate information coming from human, animal, plant, invertebrate and bacterial virology and the sciences with which they are associated – this is yet another key intermediate stop, the place occupied by the fields of comparative biology and comparative medicine.

With these thoughts about future enterprise in mind, and still with only a sense of mystery about the important selective forces, that is the causes of the evolution of viral pathogens, we are left to pure speculation in regard to the bases for the emergence of viral diseases. Given the likelihood that many selective forces are at work, and given the Pasteurian lesson in the specificity of causation that extends across all microbiology/virology, perhaps this speculation should take the form of a priority list – of course, each virologist is free to develop his/her own list.

My list is dominated by the role of humans in the emergence of new viral diseases and in the re-emergence of old diseases. Paramount, in this regard, is the ever increasing density of human, domestic animal and crop plant populations, and the consequent increased opportunity for the selection, penetrance and continued evolution of viral variants. Second, is the incredible change occurring in all ecosystems brought about by human occupation of every corner of the planet, and the consequent forced adaptation by every other species in the name of survival. This high ranking reflects my sense that most viruses representing new threats to humans are zoonotic or species jumpers and most viruses representing new threats to domestic animals and crop plants are analogous. Third, is the revolution in human movement and in the transport of things that may carry viruses, vectors and exotic hosts around the world – in these circumstances viral variants should find it easy to test new host populations and when the right niche is found to spread and evolve further to maximize "fit." Fourth, is the relative decline in global resources (and expenditures) for proven public health programs, community preventive medicine programs, agricultural extension programs, and the like, that in so many instances have stood in the way of virus spread and evolution and have been central to human well-being. The re-emergence of viral diseases that often follows such decline would seem to present fertile ground for continuing viral evolution. Fifth, is the rise of the threat of bioterrorism and biowarfare – much attention has been given to anthrax and other low-tech threats, but the high-tech genetic manipulation, the forced evolution, of viruses must not be overlooked. Sixth, is the capacity of the viruses themselves to test unoccupied niches in the absence of any apparent environtal change. Mutations leading to species jumping would seem to be most important in this regard.

Of course, this list could go on and on, but even at this point it suggests that the relationship between viral evolution and the emergence of new viral diseases is rooted in the host and the host population. In my view, we humans, as the dominant species on the planet, the only species that can affect the habitat of the viruses, might yet do more to deal with the emergence of new viral diseases – our intellect and energy has yet more to bring to the battle. So, again, what would Martinus Beijerinck, having observed the march of virologic research for 100 years, say at this point? Might he think that emerging virus diseases will soon overwhelm us, or might he look at the progress made since his seminal discoveries and think that at his bicentennial celebration spectacular new disease prevention and control successes will be reviewed and applauded? From our sense of the man and his achievements, the answer seems clear enough.

Author's address: Dr. F. A. Murphy, School of Veterinary Medicine, University of California, Davis, Davis, CA 95616-8734, U.S.A.

Life beyond eradication: veterinary viruses in basic science

L. W. Enquist

Department of Molecular Biology, Princeton University, Princeton,
New Jersey, U.S.A.

Summary. To some, the focus of research in virology entails the search for solutions of practical problems. By definition then, attention is limited to those viruses that cause disease or to exploitation of some aspect of virology to a practical end (e.g., antiviral drugs or vaccines). Once a disease is cured, or the agent eradicated, it is time to move on to something else. To others, virology offers the opportunity to study fundamental problems in biology. Work on these problems may offer no obvious practical justification; it is an affliction of the terminally curious, perhaps with the outside hope that something "useful" will come of it. To do this so-called "basic science", one must find the most tractable system to solve the problem, not the system that has "relevance" to disease. I have found that veterinary viruses offer a variety of opportunities to study relevant problems at the fundamental level. To illustrate this point, I describe some recent experiments in my laboratory using pseudorabies virus (PRV), a swine herpesvirus.

Introduction

Research in my laboratory centers on the molecular biology of neurotropic alpha-herpesviruses, a subfamily in the *Herpesviridae* family [51]. The human viruses are well known – herpes simplex virus types 1 and 2 (HSV-1, HSV-2) and varicella-zoster virus (VZV) [58]. Common domestic animals have their own unique alpha-herpesviruses as well, e.g., bovine herpes virus type 1 (BHV-1) in cattle, equine herpes virus type 1 (EHV-1) in horses, Marek's disease virus (MDV) in chickens, and pseudorabies virus (PRV) of pigs [59]. Despite having the ability to infect many cell types, these viruses invariably infect neurons in the periphery and travel inside neurons to sensory ganglia where they establish either a productive or nonproductive (latent) infection (Fig. 1). The latent infection ensures long-term survival of virus in the host population. Viral replication usually occurs first in non-neuronal cells, followed by spread of virus into afferent (e.g., sensory) or efferent (e.g., motor) nerve fibers innervating the infected tissue. Under some circumstances, virus may enter neurons directly with no prior replication in

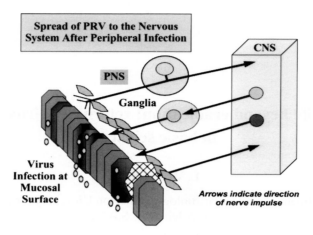

Fig. 1 Routes of spread to the nervous system after infection of a peripheral site. The cartoon illustrates virus infection at a generalized mucosal surface (left) where virus spreads among polarized epithelial cells. Depending on the surface, different cells can be found below the epithelial cell layer. Four types of neurons whose processes or cell bodies are in the periphery are illustrated. Any of these may be infected after, or concurrent with, epithelial cell infection, depending on the mucosal surface. The direction of the nerve impulse conducted by these neurons is indicated by the arrowheads. Two types of neurons whose cell bodies are in peripheral ganglia are illustrated at the top. The first is a pseudounipolar sensory neuron typical of the trigeminal ganglia whose dendrites are in the periphery and axons are in the central nervous system (*CNS*). The second is a unipolar neuron whose cell body is in the periphery (e.g., a sympathetic ganglion cell) and makes synaptic contact with axons of a neuron whose cell body is in the CNS. The third neuron type is a motor neuron with its cell body in the CNS and axon terminals in the periphery. An example would be the vagus motor neurons that innervate the viscera. The fourth neuron type has its cell body in the periphery and axon terminals in the CNS. Examples would be olfactory neurons and retinal ganglion cells. *PNS* Peripheral nervous system

non-neuronal cells; however, this is probably an infrequent event, as nerve endings are rarely exposed directly to the environment. Concurrent with the primary infection in non-neuronal cells at the epithelial surface, all alphaherpesviruses must accomplish four general processes to establish the neuronal infection: [1] virus must enter the neuron at the axon or sensory terminal, or cell body, depending on the type of neuron exposed to virus [2] the viral capsid must be transported toward the cell body of the neuron [3] viral DNA must replicate in the neuronal nucleus, and [4] virus particles must be assembled and moved out of the infected neuron in a directional manner. The direction of virus egress in the last step holds the potential for dramatically different consequences for the host. For example, following reactivation from a latent infection, virus could spread from the peripheral nervous system to the central nervous system, or it could spread back to the peripheral site serviced by that particular group of neurons. The direction taken by the virus can be the difference between a minor peripheral infection or a lethal viral encephalitis. The former is, by far, the most common outcome. Lethal CNS infection is a rare occurrence in natural hosts, but can be quite common when non-natural hosts are infected. Therefore, alphaherpesviruses must encode mechanisms to travel cell to cell in polarized epithelial cells, as well as in polarized neurons. In addition, in neurons the viruses must encode mechanisms that specify direction of virus spread within a neural circuit, but the choice of direction must be regulated.

My objectives are to understand the molecular mechanisms of herpesvirus neurotropism and spread in the mammalian nervous system. In addition, we seek to understand how the mammalian nervous system responds to neurotropic virus infections. These objectives are studied in the context of two general approaches: 1) the genetics and molecular biology of viral genes that affect virus attachment, entry, intraneuronal movement, virion assembly, transsynaptic passage and virulence and 2) the use of neurotropic viruses as tools to study the mammalian nervous system. In particular we are using these viruses as tracers of neural connections in the rodent brain and in developing chicken embryos (cf. [3, 17]).

PRV is a favorite alphaherpesvirus in my laboratory for a number of reasons. It is technically easy to work with, it grows well, purified viral DNA is highly infectious, and the virus has an amazingly broad host range. PRV can invade the nervous system and cause a lethal brain infection in diverse animals. Pigs are the natural host, but cows, dogs, cats, mice, rats, rabbits, hamsters, gerbils, Florida panthers, camels and young chickens, to name a few, can be infected [59]. I believe this virus holds answers to the fundamental questions of how invasion, spread and pathogenesis are fostered by alphaherpesviruses. Because alphaherpesviruses have conserved genes and gene functions, it makes sense to study a "veterinary" virus to learn about a human virus, and vice versa.

PRV as a tool to study alphaherpesvirus pathogenesis

All alphaherpesviruses follow a common pathway of infection; aberrations in this pathway give rise to the set of common diseases caused by these viruses (Fig. 2). Alphaherpesviruses are pantropic, infecting a wide variety of cells in culture and

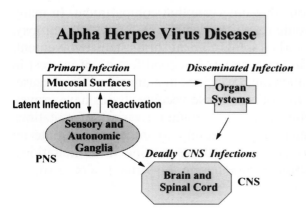

Fig. 2. An outline of the spread of alphaherpesvirus infection and its relationship to disease. *PNS* Peripheral nervous system; *CNS* central nervous system

in their hosts. Their ability to infect the peripheral and central nervous systems has attracted much attention over the years. This aspect of their life cycle results in the establishment of latent infections in the peripheral nervous system (PNS) of their hosts, a highly effective survival mechanism. Such quiescent interactions may well occur in non-neuronal cells, but few studies have addressed this issue. Variations in disease are manifest from the ability of alphaherpesviruses to spread from the PNS to the central nervous system (CNS; spinal cord and brain) or to infect peripheral tissues of non-neuronal origin.

In humans, representative diseases include epidermal lesions caused by HSV-1 and HSV-2, and chickenpox and shingles caused by VZV. Encephalitis and disseminated spread represent the most severe pathogenic results of viral infection [58]. In animals, similar pathogenic outcomes are observed after alphaherpesvirus infection, but respiratory disease, abortion, neonatal death, weight-loss and susceptibility to other microbial pathogens often are noted because of their obvious economic implications [59].

PRV is a complicated virus whose specific neurotropism and virulence are understood only in principle [43, 59]. The molecular basis for almost all aspects of CNS infection by PRV eludes us and many observations remain to be explained. Many of the genes carried by the virus have no observable function in tissue culture and therefore must function in intact animals. We cannot apply reductionist molecular biology thinking until we develop phenotypes for these genes in the animal. Once such an in vivo phenotype is found, molecular biology techniques then can be used to study function, and more tractable in vitro phenotypes can be sought. At the very least, the ability to create a genetically defined, localized infection in the mammalian CNS provides a unique window to the molecular interactions of a neurotropic parasite and its host.

Several features of the natural PRV infection have captured attention and focused interest. First, PRV infects a variety of animals, and causes lethal encephalitis in essentially all of them. The exception is the adult pig, PRV's natural host, where the virus can invade sensory ganglia and establishes a latent infection. PRV is a particularly aggressive virus in young animals, even those of its natural host, where infection is frequently fatal. One route of natural infection is via the oral-pharyngeal cavity, where virus invades mucosal epithelia and then spreads to the brain via neurons that innervate that compartment. In many young animals, infection by this route results in death in 3–5 days. At autopsy, encephalitis is often obvious, with marked spinal cord and brain stem involvement. Infectious virus is easily detected in brain tissue. A second feature of interest in PRV was the effective use of live, attenuated vaccines in managing PRV disease, thanks to the efforts of veterinary virologists. These vaccine strains have proven to be "gold mines" for viral geneticists, as they contain a variety of mutations that reduce virulence, while maintaining the ability to infect and initiate an immune response. Many of us have teased out these vaccine strain mutations, and have tried to understand their role in attenuating the virus. In doing so, we were in for some surprises.

Live, attenuated vaccine strains of PRV

Tamar Ben-Porat and colleagues performed classical studies on a PRV vaccine strain called Bartha that was developed in Eastern Europe by classical methods [4, 43]. The PRV-Bartha strain was one of several prototypes for live vaccines that subsequently provided considerable insight into the issues and problems of such agents. The PRV-Bartha strain is not a genetically engineered virus, but rather was selected after rounds of replication in tissue culture in non-swine cells. An isolate was selected that induced protective immunity in swine, yet produced few of the symptoms associated with PRV infection. PRV-Bartha strain carries a number of mutations, some of which are known, including point mutations in gC, gM, and UL21, as well as a deletion in the unique short region (U_S) (Fig. 3). It is not clear if the point mutations represent gain-of-function mutations, or if they result in reduced wild-type gene function. The deletion in the U_S region removes all, or part of the coding sequences for four genes (gI, gE, Us9 and Us2).

PRV infection traces neuronal circuitry

The use of PRV to trace neuronal circuitry in the brains of living animals may be the most fundamental, non-applied use of a veterinary virus. The identification and characterisation of synaptically-linked multineuronal pathways in the brain is important to understanding the functional organisation of neuronal circuits.

PRV Genome
142 kb

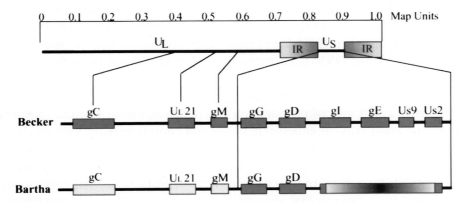

Fig. 3. Map of the PRV Becker and Bartha genomes. The map units are indicated on the first line. The general outline of the PRV genome is indicated on the second line with the unique long (U_L), unique short (U_S) and inverted repeats (*IR*) indicated. Relevant genes in the wild type strain Becker are indicated as boxes with the gene name above. Bartha contains a large deletion in the U_S region (indicated by a shaded box) that removes the gI, gE, U_S9 and U_S2 coding sequences. Bartha also contains point mutations in the gC, U_L21 and gM genes as indicated by the light color boxes. The gG and gD genes are not known to be different between the two viruses

Conventional tracing methodologies have relied on the use of markers such as wheat germ agglutinin-horseradish peroxidase, cholera toxin subunit b, fluorochrome dyes (Fluoro-Gold and Fastblue), or the plant lectin Phaseolus leucoagglutinin to delineate anterograde and retrograde pathways. The main limitations of some of these tracers are both specificity and sensitivity. During experimental manipulations, it is difficult to restrict the diffusion of certain conventional neuronal circuit tracers to a particular cell group or nucleus, so that uptake of the tracer occurs in neighboring neurons or fibers of passage and hence, false-positive labelling. Secondly, neurons located one or more synapse away from the injection site receive a progressively diminished amount of label because the tracer is diluted at each stage of transneuronal transfer. The ideal tracer then should be specific for only those connected neurons within a particular circuit, and sensitive enough to label all neurons (first-, second-, third-order, etc.) in multi-neuronal pathways. The alphaherpesviruses (PRV, HSV-1, HSV-2) have demonstrated considerable promise as self-amplifying tracers of synaptically connected neurons (see [18, 37, 39, 40] for detailed methods and reviews). Under proper conditions, second-, and third-order neurons show the same labelling intensity as first-order neurons. Moreover, the specific pattern of infected neurons observed in tracing studies are consistent with transsynaptic passage of virus, rather than lytic spread through the extracellular space (see review by Enquist et al. [18]).

Goodpasture and Teague [21, 22] initially suggested that herpes virions are taken up by nerve endings and transported to neuronal cell bodies by retrograde axonal transport. Sabin was one of the first to examine neurotropic spread of several viruses in mice, including PRV, and suggested that virus spread was consistent with known neuronal pathways [52]. Cook and Stevens [10] and Kristensson, Lycke and colleagues [30–34] were the first to demonstrate the transneuronal transfer of HSV-1 within chains of synaptically linked populations of neurons. Kuypers and co-workers [37, 57] and Dolivo and collaborators [14–16] showed the practicality of using HSV-1 and PRV, respectively, as transneuronal tracers. The use of PRV and HSV-1 as transneuronal tracers have been the subject of extensive review (see review [18]).

Both PRV and HSV-1 have similar properties as tracers. Both viruses can infect all main categories of primary sensory neurons, motor neurons, autonomic (sympathetic and parasympathetic) neurons and central nervous system sites in a variety of mammalian species. Both viruses have been used extensively to define circuits that modulate the output of both the sympathetic and parasympathetic components of the autonomic nervous system, sensory (afferent) neurons of the olfactory, optic, and trigeminal system involving the ophthalmic, maxillary, and mandibular nerve branches, as well as motor neurons of the hypoglossal, phrenic, ulnar and median nerves.

Detection of virus typically involves immunohistochemical localisation of viral antigen by light microscopy. Intracellular distribution of viral antigen is extensive and produces staining of the soma and processes of the infected cell similar to that observed using the classical Golgi method. Methodologies for the use and detection of HSV-1 or PRV for tracing neuronal connections have been extensively reviewed elsewhere [39, 40]. More recently, reporter genes have been introduced into alphaherpesvirus genomes for more direct localisation of viral infection. These include the *lacZ* gene from *E. coli* [38, 44, 54], the luciferase gene from *Photinus pyralis* [28], and the green fluorescent protein (GFP) from *Aequorea victoria* [26]. My students have constructed a variety of derivatives of the PRV-Bartha strain expressing high levels of GFP from the human cytomegalovirus promoter, as well as fusion proteins of GFP with the microtubule binding protein tau (Fig. 4). The tau-GFP fusion protein facilitates labelling of axons in brain tissue. Fusions of GFP with the Us9 protein provide brilliant markers of Golgi membranes in infected cells. Dual labelling experiments involving

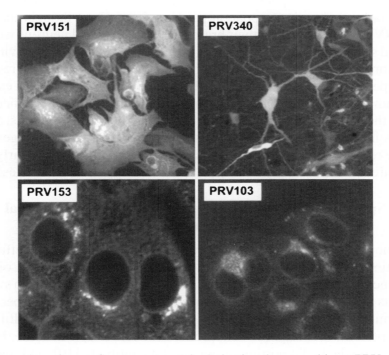

Fig. 4. Expression of green fluorescent protein derivatives by recombinant PRV strains. *PRV 151* Infection of PK15 cells. This virus expresses an enhanced GFP (EGFP) protein from the HCMV immediate early promoter. The expression cassette is inserted in the gG locus of PRV. Note that EGFP fluorescence is found throughout the cytoplasm and nucleus. *PRV 340* Infection of rat brain after heart injection with a virus that expresses a tau-EGFP fusion protein. A section of the brain stem is illustrated showing infected neurons expressing tau-EGFP. Both neuronal process and cell bodies are fluorescent. *PRV153* Infection of PK15 cells. This virus expresses a Us9-EGFP fusion protein from the HCMV immediate early promoter. The expression cassette is inserted in the gG locus. Note the intense fluorescence of the Golgi and the lack of nuclear staining. The protein is found in the virion envelope. *PRV 103* Infection of PK15 cells. This virus expresses a gE-EGFP tribrid protein comprised of the gE signal sequence fused to EGFP coding sequences, which in turn, are fused to the gE transmembrane and cytoplasmic tail coding sequences. The protein is expressed from the gE promoter. This protein concentrates in the endoplasmic reticulum, but small amounts are visible in the Golgi and plasma membrane. The protein is found in the virion envelope

the inoculation of viruses expressing different markers have been shown to identify several brain regions co-regulating multiple peripheral target organ functions [25].

The attenuated Bartha strain is an excellent tracer
of neuronal connections

Initially, my colleagues and I were surprised to find that in rodents, the attenuated strain PRV-Bartha was a far better tracer of neuronal connections than any wild type strain tested [7]. The Bartha strain infects the nervous systems of a wide variety of animals and spreads in a circuit-specific fashion whether injected into peripheral tissue, or into specific areas of the brain itself (see review [18]). Thus, we were confronted with an apparent paradox: it was long thought that animals died of PRV infection because their brains were infected and neurons were killed. Attenuated strains were thought to result from mutations that blocked virus spread and therefore blocked subsequent killing of many cells. However, the results from tracing experiments were unequivocal: the Bartha strain could spread extensively in the brains of many animals and, while the animals ultimately died of their infection, they lived several days longer, with relatively few symptoms, compared to animals infected with wild type virus. It seemed likely that viral genes, other than those responsible for neuronal infection, were involved in the early death of animals infected with wild type strains. The challenge was to find these genes.

Spread of PRV to the brain, and ability to kill an animal
do not reflect necessarily the same process

In the panoply of attenuated, live vaccine strains tested for their effectiveness in preventing PRV disease, many scientists have found that deletion of the non-essential membrane protein gE is critical for attenuation of the virus in swine [24]. gE mutants also exhibit reduced virulence in almost every animal species tested that is permissive for PRV [24, 59]. The PRV gE gene encodes a multifunctional virulence protein; it is required for spread from cell to cell in some, but not all cell types, it binds the Fc portion of porcine IgG antibody [19], and it appears to have intrinsic virulence properties.

The intrinsic virulence properties of gE can be seen easily when PRV spreads to the brain by retrograde infection of a cranial nerve after infection at the periphery. As an example, we focused on the tenth cranial nerve (the vagus nerve), which innervates the laryngeal mucosa, the oesophagus, the thorax, and most of the abdomen. We took advantage of the well established innervation of the stomach musculature by the vagus, to follow retrograde spread of wild type and attenuated strains of PRV from axon terminals in the stomach muscles, to the cell bodies of the primary neurons of the vagus nerve in the brain stem, and then to synaptically connected neurons throughout the brain. We injected approximately 10^5 plaque forming units of each of five virus strains into the stomach muscles of different rats and followed spread of each virus into the nervous system. The strains used were wild type Becker strain, the Bartha vaccine strain, plus three gE

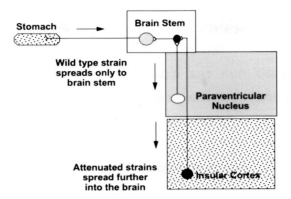

Fig. 5 Spread of PRV In the vagus circuitry after stomach muscle injection. The cell bodies of motor neurons that project axons to the stomach muscles are located in the dorsal motor vagal nucleus (DMV) of the brain stem. Neurons in the nucleus of the solitary tract (NTS) lie immediately above the dorsal motor vagal nucleus and are in synaptic contact with DMV neurons. Neurons in the paraventricular nucleus (PVN) (around the third ventricle) project axons back to synapse on NTS neurons. Neurons in the insular cortex (IC) in the forebrain also project axons back to synapse on NTS neurons. The wild type virus spreads from the stomach to the brain stem infecting the DMV and NTS, but animals die before further spread to the PVN and IC can occur. Animals survive longer after infection with Bartha or Becker gE mutants and spread further in the vagus circuit

mutants isogenic with the Becker strain. PRV 25 contains a frame-shift mutation affecting the cytoplasmic tail of gE, PRV 26 contains a nonsense mutation just before the gE transmembrane domain such that the truncated gE protein is secreted from the cell, and PRV 91 carries a deletion of the entire coding sequence of gE. Serial, coronal sections through the brains of infected rats killed at the indicated times were stained for viral structural proteins with a polyvalent antisera and a peroxidase-linked secondary antibody.

The results were quite striking: wild type Becker virus killed all animals by 60 hours after infection, but the virus had travelled essentially to the brain stem, and no further. The Bartha strain, in contrast, spread through second- and third-order synaptic connections infecting the extent of the known vagus circuitry, traveling to the insular cortex and beyond. Animals sustained this massive brain infection for at least five to six days after infection with little or no symptomatology. A single point mutation in gE (e.g., PRV 25 or PRV 26) or a deletion of gE (PRV 91) in the virulent Becker strain, facilitated spread of virus in the vagus circuitry such that these viruses spread almost as far in the brain as the Bartha strain (Fig. 5). This remarkable finding is evidence that the gE protein is a virulence factor, playing a role in causing animals to die. However, this function is not related to the ability of these viruses to infect the nervous system. It may appear to be counterintuitive, but gE mutants spread further and infect more cells in the brain than their virulent parent because the animals survive longer. Moreover, as suggested by the gE mutant viruses PRV 25 and PRV 26, the cytoplasmic tail of gE is critical for expression of this early virulence function [55].

Cytoplasmic Tail of PRV gE

R [TM] RRR AAS RPF RVP TRA GTR

MLS PV*Y TSL* PTH EDY YDG DDD DEE

AGD ARR RPS SPG GDS GYE GP*Y VSL*

DAE DEF SSD EDD GLY VRP EEA PRS

GFD VWF RDP EKP EVT NGP NYG VTA

SRL LNA RPA -COOH

Fig. 6. The amino acid sequence of the PRV Becker gE cytoplasmic tail. Amino acids are indicated by the single letter code. The transmembrane region is noted by "TM" in a box at the beginning of the sequence. The putative endocytosis signals (general motif YXXL) are underlined and in italics. Potential casein kinase II phosphorylation of two serines is indicated by the boxed SS sequence

The gE cytoplasmic tail: endocytosis functions

In the last two years, several laboratories reported that certain VZV envelope proteins did not remain on the surface of cells that expressed them, but rather were internalised by endocytosis in a recycling process [48, 62]. The biological function of membrane protein endocytosis in the virus life cycle remains a matter of speculation and debate. My students and I have demonstrated that some, but not all, membrane proteins encoded by PRV internalise after reaching the plasma membrane [56]. Glycoproteins gE and gB internalise from the plasma membrane of cells, while gI and gC do not internalise efficiently. We have found that the cytoplasmic domain of gE is required for its internalisation. Indeed, two endocytosis motifs, of the general sequence YXXL, are found in the gE tail (Fig. 6). PRV gE internalises from the plasma membrane in the absence of other viral proteins and also directs endocytosis of its binding partner, gI. During infection, internalisation of the gE/gI complex is inhibited after 6 h of infection (Fig. 7). To test the role of endocytosis in the viral life cycle, we have engineered

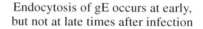

Endocytosis of gE occurs at early, but not at late times after infection

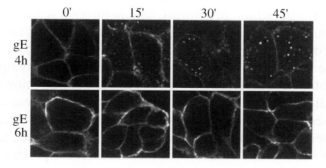

Fig. 7. Endocytosis of PRV gE at 4 h and 6 h after infection. The experiment was done as described by Tirabassi et al. [56]. At the indicated time after infection, cells were put on ice and exposed to a monoclonal antibody that recognized the gE/gI complex. Cells were then shifted to 37 C to initiate endocytosis. At the indicated times, cells were fixed, permeabilized and the gE/gI-specific antibody detected by an FITC conjugated second antibody

gE Tail Mutant Phenotypes

gE	Endocytosis	Virion	Virulence	Plaque size, MDBK
WT	Yes	Yes	+++	++++
Y517S	Yes	Yes	n.d.	++++
Y478S	Inhibited	Yes	+++	++
Double	Inhibited	Yes	+++	++
Tail-anc	Inhibited	Yes	+/−	++
Tail-fs	Inhibited	No	+/−	+
Tail-sec	Inhibited	No	+/−	++
Null	——	No	−	+

Fig. 8. Summary of phenotypes of PRV gE mutants. The mutations are listed in the first column. *WT* is the wild type PRV Becker gE protein. *Y517S* is the gE mutant protein where the tyrosine at position 517 is mutated to a serine. This mutation affects the second YXXL putative endocytosis motif. *Y478S* is a similar mutation in the first YXXL putative endocytosis motif. *Double* is a double mutation of Y517S and Y478S. Both YXXL endocytosis motifs are mutated. *Tail-anc* is a mutant gE protein that is anchored in the membrane but lacks a cytoplasmic tail due to a nonsense mutation at codon 457. *Tail-fs* is a mutant gE protein that is anchored in the membrane but has a frame-shift mutation at codon 446 such that the transmembrane domain is 7 amino acids shorter than wild type and novel amino acids replace the cytoplasmic domain. *Tail-sec* is a mutant gE protein that lacks the transmembrane and cytoplasmic domain; it is secreted into the medium. The column labeled *endocytosis* indicates that the gE protein is internalised (yes) or stays at the cell surface (inhibited) at 4 h after infection. The column labeled *virion* indicates if the mutant gE protein is found in purified virions. The column labeled *virulence* scores the ability of the indicated virus to cause disease in rats after retinal infection by 10^5 plaque forming units of virus. N.D. means not determined. +++ means disease caused by wild type virus. +/− means significant reduction in disease, but more than observed for the gE null mutant (−). *Plaque size, MDBK* describes the size of plaques on MDBK cells. Wild type plaques are large (++++) and gE null mutant plaques are minute (+). Plaques of intermediate size are scored ++. The boxed scores highlight the small plaque phenotype of Y478S, the double endocytosis mutant and the tail-anchored mutant

viral mutants encoding mutations in two putative internalisation signals (YXXL) in the gE cytoplasmic tail. A mutation in the first or both motifs decreases internalisation of the gE/gI complex, while gE protein with a mutation in the second motif internalises as efficiently as, or better than, wild-type gE. To test if the gE cytoplasmic tail contains all signals necessary to direct internalisation of the protein, we have constructed a virus encoding a hybrid GFP-gE tail fusion (see Fig. 4, PRV 103). The gE transmembrane domain and cytoplasmic tail is sufficient to target the GFP molecule into the secretory pathway as well as into virions. We are testing whether the gE tail also directs internalisation of the GFP molecule. A summary of the phenotypes of these mutants is given in Fig. 8.

We do not yet understand the function of gE endocytosis in the PRV life cycle. One idea was that internalisation was required for insertion of gE into a virion envelope. In this model, the virus would acquire its envelope from a late endosomal compartment that would receive mature viral membrane proteins by endocytosis. Our data suggest that this idea is not correct for PRV. We have shown that gE protein internalised at 4 h post infection is not present in virions formed at a later time [56]. More specifically, we have shown recently that viruses with mutations in the gE internalisation signal incorporate gE efficiently in virion envelopes. Another idea was that endocytosis was required for expression of virulence. This also is not likely as mutants defective in gE endocytosis remain as virulent as wild type virus. Surprisingly, endocytosis mutants have defects in cell-cell spread as measured by plaque size on tissue culture cells, but appear to spread like wild type virus in the rodent CNS. These findings again remind us that spread from cell to cell measured in tissue culture is not always correlated with virulence.

US9, a type II, tail-anchored membrane protein

The Us9 gene is deleted in the Bartha vaccine strain and attracted our attention because Us9 is highly conserved among the alphaherpesviruses sequenced to date, including VZV, which has the smallest alphaherpesvirus genome. The HSV-1 Us9 homologue was reported to be a tegument protein and to be associated with nucleocapsids in the nuclei of infected cells [20]. This assignment as a tegument protein is commonly made for all Us9 homologues. We have determined that the PRV Us9 protein is considerably different: it is a novel type II membrane protein that is highly phosphorylated [5]. It localizes to the secretory system (predominately to the Golgi apparatus) and not to the nucleus. By fusing the jellyfish enhanced green fluorescent protein reporter molecule (EGFP) to the carboxy-terminus of Us9, we demonstrated that Us9 not only is capable of targeting a Us9-EGFP fusion protein to the Golgi compartment (see Fig. 4, PRV 153), it also is able to direct efficient incorporation of such chimeric molecules into infectious viral particles. The predominant localisation of Us9 to the Golgi apparatus may have important ramifications for models of herpesvirus envelopment. The Us9 protein lacks a signal sequence and is probably inserted in membranes post-translationally. To the extent of our knowledge, this is the first identification of a type II, tail anchored membrane protein in alphaherpesvirus envelopes.

We have used deletion analysis to study the signals required for localisation of Us9 to the Golgi apparatus and for incorporation into the viral envelope. Preliminary results indicate that a highly conserved region containing potential tyrosine and casein kinase I and II phosphorylation sites is important for the localisation of Us9 to the Golgi apparatus. Deletion of these 10 amino acids resulted in relocalisation of Us9 to the plasma membrane in both transfected and infected cells. The deletion of this region, however, had no effect on incorporation of Us9 into the viral envelope. In addition, preliminary experiments on a cell line stably expressing a Us9-GFP fusion protein suggest that wild type Us9 molecules not only are able to travel to the cell surface, but also those molecules that do reach the plasma membrane are subsequently internalised. The newly internalised molecules return to a cellular compartment reminiscent of the Golgi where they

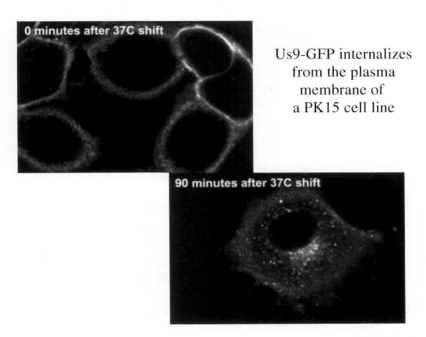

Fig. 9. Endocytosis of Us9-GFP. The PK15 cell line expressing a Us9-GFP fusion protein is described in [5]. At zero time (top panel), the cells were placed on ice and exposed to a monoclonal antibody specific for GFP which will react only with GFP present on the cell surface. The cells then are shifted to 37 C, and at 90 min, the cells are fixed and permeabilized. The GFP antibody (and Us9-GFP protein) are localized with a Cy-3 conjugated second antibody (red). Green GFP fluorescence is not visible

closely associate with non-internalised Us9 molecules (Fig. 9). One idea is that the Us9 protein is maintained at steady state in this compartment by efficient retrieval from the plasma membrane. The Us9 protein does not have a YXXL internalisation motif as do gE and gB, so the signals by which it internalises remain to be discovered. The roles of US9 internalisation and Golgi localisation in the virus life cycle are under study. A Us9 null PRV strain has been created and preliminary results indicate that the mutant has defects in virulence and spread in the rat nervous system.

Allele-specific regulation of MHC class I by pseudorabies virus

Identification of viral genes that influence host defences has become an important part of research in viral pathogenesis. Many herpesviruses have been shown to modulate major histocompatibility complex class I (MHC-I) expression including HSV, Human and murine cytomegalovirus, BHV-I, VZV and PRV [9, 23, 27, 41, 47]. The molecular mechanism by which PRV regulates MHC I is unclear, as the PRV genome contains no obvious gene homologues related to known herpesvirus genes that affect MHC expression. Thus, this virus may offer the opportunity to find new molecules that interact with the immune system. We have measured MHC-I expression on the surface of PRV-infected mouse

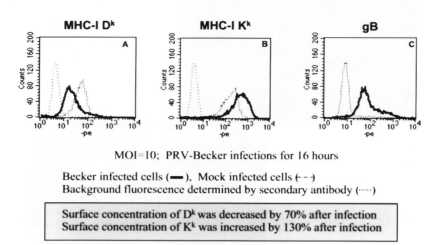

Fig. 10. FACScan of MHC-I expression on the cell surface of murine L929 cells after PRV infection. L929 cells were infected at an MOI of 10. At 16 h post-infection, cells were fixed and stained for MHC-I with monoclonal antibodies specific for the Kk or Dk allele (**A** and **B**). Samples were also stained for PRV gB with a gB specific antibody (**C**). This measures the extent of virus infection in the population. The signal from infected cells is indicated by the dark solid line. The signal from mock infected cells is indicated by the dashed line. The signal from secondary antibody alone is indicated in **A** by the far left curve

fibroblasts (L929) and have shown that MHC-I cell surface expression is regulated in an allele specific manner such that cell surface expression of the K^k allele is increased 130% during infection and the D^k allele is decreased 70% (Fig. 10). By comparing MHC-I expression using the wild type strain Becker and the vaccine strain Bartha, we have shown that D^k and K^k are regulated by two distinct mechanisms that can be separated genetically, temporally, and mechanistically. In swine cells, we have shown that PRV increases MHC I in a B cell line (L14) and decreases MHC-I in swine kidney cells (PK 15). By using low pH citrate washes to remove MHC-I from the cell surface, we have also shown that the early decrease of D^k at 4 h post infection is not due to a block in synthesis or transport of MHC-I to the cell surface. In contrast, later in infection, at 8 h post infection, cell surface expression of both alleles is inhibited. These results suggest that PRV contains genes that can both increase and decrease cell surface expression of MHC-I. Finding and characterising these genes may shed light, not only on how PRV interacts with the host immune system, but may also provide insight into the mechanisms functioning in the MHC-I pathway.

PRV BACs

An unbiased screen for virus mutants that affect pathogenesis has been difficult to perform. We and others have used attenuated, live vaccine strains to find genes that influence viral pathogenesis. In addition, targeted mutagenesis of known herpesvirus genes has been another method to assess viral gene functions in pathogenesis and viral replication. Traditional methods rely on homologous recombination of purified viral DNA with an engineered plasmid following co-transfection in mammalian cells. This can be an inefficient process that often requires cumbersome screening and plaque purification steps. We are exploring alternative approaches to genetic manipulation of PRV that largely circumvent these problems. These methods center around construction and manipulation of an infectious clone of the full-length viral genome ($\sim$142kbp) in a mini-F plasmid (a bacterial artificial chromosome or BAC, [42]). Mutagenesis of one PRV-Becker infectious clone in *E. coli* by the methods of allelic exchange and transposon mutagenesis is underway (Fig. 11). Currently, we have constructed several variations

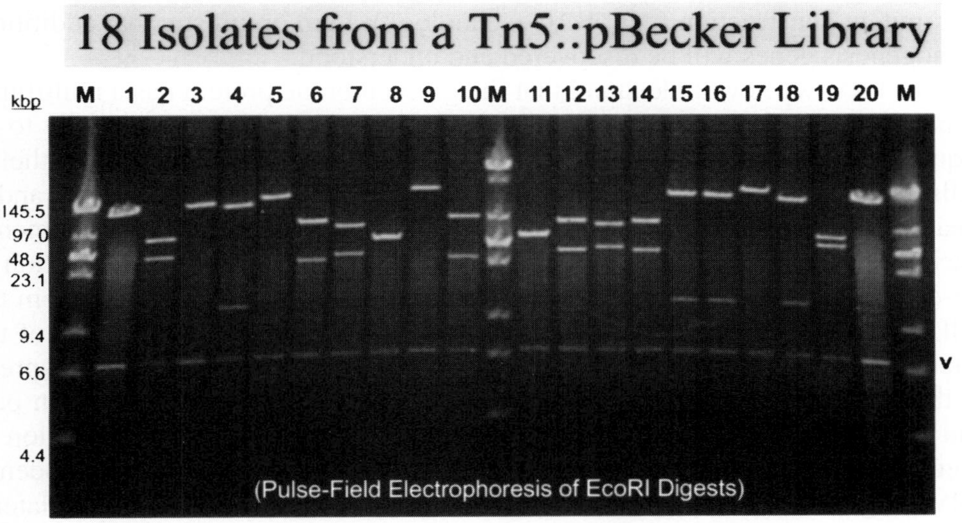

Fig. 11. EcoRI restriction enzyme analyses of PRV-BAC plasmids with Tn5 transposon insertions. The PRV-Becker BAC plasmid contains the approximately 7 Kb F replicon harboring a marker for chloramphenicol resistance in the gG locus of PRV. The F replicon also contains two Eco RI sites near each junction with PRV DNA. The Tn5 transposon contains a single EcoRI site, while the PRV Becker genome has no EcoRI sites. The PRV Becker BAC plasmid was transformed in to *E. coli* cells containing the Tn5 transposon (carrying a kanamycin resistance marker) in a conditional growth plasmid, and grown for several generations. The culture was then transformed to another *E. coli* cell where only the BAC plasmid would replicate and the culture grown in the presence of kanamycin and chloramphenicol to select for BACs with Tn5 insertions. Plasmid DNA was purified and transformed into *E. coli* and single clones picked for DNA analysis after growth on kanamycin/chloramphenicol containing LB agar. Plasmid DNA from 18 single colonies was digested with EcoRI and fractionated by pulse-field gel electrophoresis. Of the 18 isolates shown here, 11 were unique and 7 were probably derived from the same transposition event. Four isolates contained insertions of the transposon in the U_s region and 7 had insertions in the U_L region. The lane marked *M* contains molecular size markers. *1* and *20* contain the EcoRI digested PRV-BAC plasmid with no transposon insertion. Note the intact 142 Kb PRV genome and the approximately 7 Kb plasmid replicon (*v*)

of PRV infectious clones and are analysing several recombinant mutant viruses produced by these methods.

Conclusions and questions

Veterinary viruses have much to offer in our quest to understand the basic principles governing the neurotropism and virulence of alphaherpesviruses. While many laboratories are approaching these questions from different perspectives using a variety of viruses, we all have only begun to define the problems, much less to uncover the molecular mechanisms. As demonstrated by several laboratories, the approach of studying attenuated PRV vaccine strains has provided much insight into the issues involved in alphaherpesvirus pathogenesis. However, we must go further to find additional pathogenesis genes. When one couples the ability to work with natural host infections, the ability to establish many animal models due to the broad host range of the virus, with new methods in mutagenesis and mutant construction (e.g., BAC technology), I am optimistic that additional pathogenesis genes will be discovered and understood.

For PRV, we now understand that the gE membrane protein has a multifunctional role in the viral life cycle and pathogenesis [24]. First, gE is likely to be required for efficient cell-to-cell spread at the site of primary infection of epithelial cells. Without gE, it is likely that a primary infection is poorly established, and is cleared rapidly by host innate defences (e.g., interferons, NK cells, complement). Lack of this cell-spread function probably is the primary basis for attenuation of gE-deleted viruses. However, gE often is required for transfer of virus from the cell body of neurons to neurons in synaptic contact with axon terminals of the infected cell (anterograde spread). In pigs, gE/gI is required for PRV to spread to the olfactory bulb after infection of the nasal olfactory mucosa, and in pigs and rodents, to the trigeminal motor nuclei in the brain stem after infection of pseudounipolar sensory neurons of the 5th cranial nerve [2, 35, 36]. In rodents, PRV gE/gI are required for anterograde infection of the optic tectum and lateral geniculate, but not the suprachiasmatic nucleus [8, 60]. In the rodent CNS, gE/gI are required, in part for anterograde transport from the prefrontal cortex to the striatum [6]. This function of gE also may be required for directional spread of newly produced virus to the mucosal surface after reactivation from latency. Finally, gE apparently has intrinsic virulence functions that are independent of its role in cell-cell spread; such functions require the cytoplasmic tail. At least two functions are known to be encoded in this part of the protein: endocytosis and serine phosphorylation [24, 55, 56]. Endocytosis has no obvious role in the expression of virulence. The role of phosphorylation remains to be explored. One hypothesis is that on the surface of an infected cell, gE binds a ligand (perhaps an antibody-like molecule) that activates a signal-transduction cascade in the infected cell in a process requiring the gE cytoplasmic tail. As a result of this signaling, the infected cell would produce interferons or other biological modifiers that affect virulence. As the gE tail is phosphorylated and the site of phosphorylation is known, we are in a position to test the first order predictions of this hypothesis directly.

In searching for the virulence functions encoded in the gE cytoplasmic tail, we found endocytosis signals that recycle the protein from its position on the plasma membrane of infected cells. As noted above, these signals apparently have no function in virulence, but are involved in cell-cell spread in cultured cell lines. Fundamental questions remain. Why do some, but not all alphaherpesvirus envelope membrane proteins internalise from the plasma membrane? Why does endocytosis of some viral proteins occur early, but not late after infection? Why is membrane protein endocytosis shut off at all in infected cells? The answers to these questions await further study.

The role of the small type II membrane protein called Us9 in the biology of the alphaherpesviruses is not understood. It is a major component of the virion envelope, it is phosphorylated, it localises to a Golgi compartment in infected cells by retrieval from the plasma membrane, and it is conserved in every human and veterinary virus studied. Viruses lacking this gene have few obvious phenotypes in infected tissue culture cells. This protein must play a role in the survival of alphaherpes viruses in nature, but its function still eludes us. The structure of the Us9 protein is similar to SNARE proteins such as synaptobrevin involved in vesicle targeting and fusion. Perhaps, like synaptobrevin, PRV Us9 protein might be translated in the cytoplasm and post-translationally inserted directly into secretory membranes. If this speculation can be substantiated, it would challenge some of the models for envelopment of herpesviruses.

One major mystery concerns the mechanism(s) by which infectious virus is transferred from one infected neuron to an uninfected neuron (see review [18]). If there is one take home message learned from studies with our neuroanatomy colleagues who use PRV and HSV as tracers of neural connections, it is that with rare exception, the viruses travel from neuron to neuron in functional neuronal circuits, as defined by classical neuroanatomy and physiology. They rarely spread non-specifically, even when injected directly into the CNS, or when they infect sites never infected by the virus naturally. One way that circuit-specific infection could occur is if HSV and PRV leave neurons at, or near, sites of synaptic contact such that virus is taken up by connected cells. This assertion raises three general unanswered questions: what is the nature of the virus particle being transferred to uninfected neurons, how is this intracellular virus particle targeted to sites of synaptic contact during viral egress, and what is the mechanism(s) for trans-synaptic passage? Transport of virus out of an infected cell is intimately involved with the processes of virion assembly, a process that is not yet understood in detail for any herpesvirus. It seems clear that movement of virions or subvirion particles inside a cell must be governed by the basic cell biology of neurons and use the motor proteins found in particular compartments. For example, this might mean that transport of virus toward the nucleus would use dynein-type motors in axons, and transport of virus away from the nucleus would use kinesin-type motors. Several studies indicate that viral gene products are involved in directional spread of PRV [1, 2, 8, 35, 36] and HSV [13, 61] in the nervous system, but the search for such genes has not been exhaustive. The BAC system as outlined above (see also [42]) may enable us to perform unbiased genetic screens for PRV mutants that affect the many steps in this process.

In thinking about how virus infects the synaptically connected, uninfected neuron, we should be wary of setting artificial boundaries based on data obtained from cultured, non-neuronal cells. For example, in the simplest case for transneuronal viral infection, all that needs to be transferred to the uninfected neuron is the viral genome. A completed, mature virion need not be the vehicle for transfer of genomes and tegument proteins between neurons. Immature enveloped virions, naked capsids or even individual proteins could be transferred to neurons at or near sites of synaptic contact. Additionally, the processes of exocytosis and endocytosis are active at synapses and there is no reason why they could not be used by viruses. Mature extracellular virions, as identified in tissue culture experiments, may be required primarily for passage between hosts or for infection of mucosal epithelia. Evidence for this speculation comes from work in PRV where the gD protein is required for mature virion infectivity, but not for cell-cell spread in tissue culture (plaque formation) or for neuron-to-neuron spread in animals [1, 46, 49, 50, 53]. This is not true for HSV, where gD is required for every known mode of virion and cell-cell infection [12]. It is interesting to note that VZV does not have a gD gene implying that in this virus, other viral proteins provide gD function [11]. However, both HSV and PRV require the gB and gH/gL gene products for infection and spread in neurons of every model tested (anterograde or retrograde spread) [1, 2]. As these proteins are thought to trigger pH-independent fusion of the viral envelope with the host cell plasma membrane, it is likely that such an event must be inherent in neuron-to-neuron spread.

Finally, little can be said at this time regarding the role of the host immune system and the function of specific viral receptors in neurotropism. It is our hope that neurotropic viruses like PRV can be used to dissect the immune defenses of the nervous system, including regulation of MHC expression. It is apparent that the alphaherpesviruses have mechanisms to help them establish a primary infection at mucosal surfaces, mechanisms that facilitate their survival when reactivation of a latent infection occurs in an immunized animal, and mechanisms that facilitate their entry into and passage through the peripheral and central nervous system. At this writing, no evidence exists for a unique neuronal receptor, but rather for many co-receptors. The herpesvirus entry mediator (HVEM; now called HveA) that facilitates entry of many HSV strains to lymphocytes, is not the only receptor used by HSV and PRV [45]. Other alphaherpesvirus receptors are being identified, and we may soon have sufficient information to tell us if any of these herpes receptors function only in the nervous system. Despite many years of work and intensive study, we clearly have much to learn about the molecular mechanisms of spread and pathogenesis of alphaherpesviruses in the nervous system. Nevertheless, I believe that veterinary viruses such as PRV have had, and will continue to have, significant impact in this process.

Acknowledgements

I thank the organisers of the Greifswald Symposium, "One hundred years of virology – past, present and future of virus research", for the opportunity to speak and summarise our work. I thank my colleagues J. Patrick Card and Richard Miselis for advice, help and insight into rodent neuroanatomy and virus infection. Members of the PRV research community have always provided reagents, advice and healthy competition, for which I am grateful. My secretary Trisha Barney and my technician Marlies Eldridge have kept me and the lab organised. Finally, I wish to acknowledge the members of my laboratory, past and present, who have done all of the work discussed in this report. Our work is supported by grants from the National Institutes of Neurological Diseases and Stroke.

Note added in proof

References [63]–[67] – publications of work described in this review – were added after submission of the paper.

References

1. Babic N, Mettenleiter TC, Flamand A, Ugolini G (1993) Role of essential glycoproteins gII and gp50 in transneuronal transfer of pseudorabies virus from the hypoglossal nerves of mice. J Virol 67: 4 421–4 426
2. Babic N, Klupp B, Brack A, Mettenleiter TC, Ugolini G, Flamand A (1996) Deletion of glycoprotein gE reduces the propagation of pseudorabies virus in the nervous system of mice after intranasal inoculation. Virology 219: 279–284
3. Banfield BW, Yap GS, Knapp AC, Enquist LW (1998) A chicken embryo eye model for the analysis of alphaherpesvirus neuronal spread and virulence. J Virol 72: 4 580–4 588
4. Bartha A (1961) Experimental reduction of virulence of Aujeszky's disease virus. Magy Allatorv Lapja 16: 42–45
5. Brideau AD, Banfield BW, Enquist LW (1998) The Us9 gene product of pseudorabies virus, an alpha herpesvirus, is a phosphorylated, tail anchored type II membrane protein. J Virol 72: 4 560–4 570
6. Card JP, Levitt P, Enquist LW (1998) Different patterns of neuronal infection after intracerebral injection of two strains of pseudorabies virus. J Virol 72: 4 434–4 441
7. Card JP, Rinaman L, Schwaber JS, Miselis RR, Whealy ME, Robbins AK, Enquist LW (1990) Neurotropic properties of pseudorabies virus: uptake and transneuronal passage in the rat central nervous system. J Neurosci 10: 1 974–1 994
8. Card JP, Whealy ME, Robbins AK, Moore RY, Enquist LW (1991) Two α-herpesvirus strains are transported differentially in the rodent visual system. Neuron 6: 957–969
9. Cohen J (1998) Infection of cells with varicella-zoster virus down-regulates surface expression of class I major histocompatibility complex antigens. Infect Dis 155: 1 390–1 393
10. Cook ML, Stevens JG (1973) Pathogenesis of herpetic neuritis and ganglionitis in mice: evidence for intra-axonal transport of infection. Infect Immun 7: 272–288
11. Davison AJ, Scott JE (1986) The complete DNA sequence of varicella-zoster virus. J Gen Virol 67: 1 759–1 816
12. Dingwell KS, Brunetti CR, Hendricks RL, Tang Q, Tang M, Rainbow AJ, Johnson DC (1994) Herpes simplex virus glycoproteins E and I facilitate cell-to-cell spread in vivo and across junctions of cultured cells. J Virol 68: 834–845

13. Dingwell KS, Doering LC, Johnson DC (1995) Glycoproteins E and I facilitate neuron-to-neuron spread of herpes simplex virus. J Virol 69: 7 087–7 098

14. Dolivo M (1980) A neurobiological approach to neurotropic viruses. Trends Neurosci 3: 149–152

15. Dolivo M, Beretta E, Bonifas V, Foroglou C (1978) Ultrastructure and function in sympathetic ganglia isolated from rats infected with pseudorabies virus. Brain Res 140: 111–123

16. Dolivo M, Honegger P, George C, Kiraly M, Bommeli W (1979) Enzymatic activity, ultrastructure and function in ganglia infected with a neurotropic virus. In: Cuenod M, Kreutzberg GW, Bloom FE (eds) Development and chemical specificity of neurons, Vol 51. Elsevier/North-Holland Biomedical Press, Amsterdam, pp 51–57

17. Enquist LW, Card JP (1996) Pseudorabies virus: a tool for tracing neuronal connections. In: Lowenstein R, Enquist LW (eds) Protocols for gene transfer in neuroscience: towards gene therapy of neurological disorders. J Wiley, Chichester, pp 333–348

18. Enquist LW, Husak PJ, Banfield BW, Smith GA (1998) Infection and spread of alphaherpesviruses in the nervous system. Adv Virus Res 58: 237–347

19. Favoreel HW, Nauwynck JJ, van Oostveldt P, Mettenleiter TC, Pensaert MB (1997) Antibody-induced and cytoskeleton-mediated redistribution and shedding of viral glycoproteins, expressed on pseudorabies virus-infected cells. J Virol 71: 8 254–8 261

20. Frame MC, McGeoch DJ, Rixon FJ, Orr AC, Marsden HS (1986) The 10 K virion phophoprotein encoded by gene Us9 from herpes simplex virus type I. Virology 150: 321–332

21. Goodpasture EW, Teague O (1923) Transmission of the virus of herpes fibrils along nerves in experimentally infected rabbits. J Med Res 44: 139–184

22. Goodpasture EW (1925) The axis-cylinders of peripheral nerves as portals of entry to the central nervous system for the virus of herpes simplex in experimentally infected rabbits. Am J Pathol 1: 11–33

23. Hengel H, Koszinowski UH (1997) Interference with antigen processing by viruses. Curr Opin Immunol 9: 470–476

24. Jacobs L (1994) Glycoprotein E of pseudorabies virus and homologous proteins in other alphaherpesvirinae. Arch Virol 137: 209–228

25. Jansen ASP, Nguyen XV, Karpitskiy V, Mettenleiter TC, Loewy AD (1995) Central command neurons of the sympathetic nervous system: basis of the fight-or-flight response. Science 270: 644–646

26. Jons A, Mettenleiter TC (1997) Green fluorescent protein expressed by recombinant pseudorabies virus as an in vivo marker for viral replication. J Virol Methods 66: 283–292

27. Kimman TG, Bionchi ATJ, de Bruin TGM, Mulder WAM, Priem J, Voermans JM (1995) Interaction of pseudorabies virus with immortalized porcine B cells: influence on surface class I and II major histocompatibility and immunoglobulin M expression. Vet Immunol Immunopath 45: 253–263

28. Kovacs SzF, Mettenleiter TC (1991) Firefly luciferase as a marker for herpesvirus (pseudorabies virus) replication in vitro and in vivo. J Gen Virol 72: 2 999–3 008

29. Kristensson K (1970) Morphological studies of the neural spread of herpes simplex virus to the central nervous system. Acta Neuropathol 16: 54–63

30. Kristensson K, Ghetti B, Wisniewski HM (1974) Study on the propagation of herpes simplex virus (type 2) into the brain after intraocular injection. Brain Res 69: 189–201

31. Kristensson K, Lycke E, Sjostrand J (1971) Spread of herpes simplex virus in peripheral nerves. Acta Neuropathol 17: 44–53

32. (Deleted in proof)

33. Kristensson K, Nennesmo I, Persson L, Lycke E (1982) Neuron to neuron transmission of herpes simplex virus: transport of virus from skin to brainstem nuclei. J Neurol Sci 54: 149–156

34. Kristensson K, Vahlne A, Persson LA, Lycke E (1978) Neural spread of herpes simplex virus types 1 and 2 in mice after corneal or subcutaneous (footpad) inoculation. J Neurol Sci 35: 331–340

35. Kritas SK, Pensaert MB, Mettenleiter TC (1994) Role of envelope glycoproteins gI, gp63 and gIII in the invasion and spread of Aujeszky's disease virus in the olfactory nervous pathway of the pig. J Gen Virol 75: 2 319–2 327

36. Kritas SK, Nauwynck HJ, Pensaert MB (1995) Dissemination of wild type and gC-, gE- and gI-deleted mutants of Aujeszky's disease virus in the maxillary nerve and trigeminal ganglion of pigs after intranasal inoculation. J Gen Virol 76: 2 063–2 066

37. Kuypers HGJM, Ugolini G (1990) Viruses as transneuronal tracers. Trends Neurosci 13: 71–75

38. Loewy AD, Bridgman PC, Mettenleiter TC (1991) Beta-galactosidase expressing recombinant pseudorabies virus for light and electron microscopic study of transneuronally labeled CNS neurons. Brain Res 555: 346–352

39. Loewy AD (1995) Pseudorabies virus: a tranneuronal tracer for neuroanatomical studies. In: Kaplitt MG, Loewy AD (eds) Viral vectors. Gene therapy and neuroscience applications. Academic Press, San Diego, pp 349–366

40. Lowenstein PR, Enquist LW (1995) Protocols for gene transfer in neuroscience; towards gene therapy of neurological disorders. J Wiley, New York

41. Mellencamp MW, O'Brien PCM, Stevenson JR (1991) Pseudorabies virus-induced suppression of major histocompatibility complex class I antigen expression. J Virol 65: 3 365–3 368

42. Messerle M, Crnkovic I, Hammerschmidt W, Ziegler H, Koszinowski UH (1997) Cloning and mutagenesis of a herpesvirus genome as an infectious bacterial artificial chromosome. Proc Natl Acad Sci USA 94: 14 759–14 763

43. Mettenleiter TC (1994) Pseudorabies (Aujeszky's Disease) virus: state of the art. Acta Vet Hung 42: 153–177

44. Mettenleiter TC, Rauh I (1990) A glycoprotein gX-β-galactosidase fusion gene as insertional marker for rapid identification of pseudorabies virus mutants. J Virol Methods 30: 55–66

45. Montgomery RI, Warner MS, Lum BJ, Spear PG (1996) Herpes simplex virus-1 entry into cells mediated by a novel member of the TNF/NGF receptor family. Cell 87: 427–436

46. Mulder W, Pol J, Kimman T, Kok G, Priem J, Peeters B (1996). Glycoprotein D-negative pseudorabies virus can spread transneuronally via direct neuron-to-neuron transmission in its natural host, the pig, but not after additional inactivation of gE or gI. J Virol 70: 2 191–2 200

47. Nataraj C, Eidmann S, Hariharan MJ, Sur JH, Perry GA, Srikumaran S (1997) Bovine herpesvirus 1 downregulates the expression of bovine MHC class I molecules. Viral Immunol 10: 21–34

48. Olson JK, Grose C (1997) Endocytosis and recycling of varicella-zoster virus Fc receptor glycoprotein gE: internalisation mediated by a YXXL motif in the cytoplasmic tail. J Virol 71: 4 042–4 054

49. Peeters B, Pol J, Gielkens A, Moormann R (1993) Envelope glycoprotein gp50, of

pseudorabies virus is essential for virus entry but is not required for viral spread in mice. J Virol 67: 170–177

50. Rauh I, Mettenleiter TC (1991) Pseudorabies virus glycoproteins gII and gp50 are essential for virus penetration. J Virol 65: 5 348–56

51. Roizman B, Sears E (1996) Herpes simplex viruses and their replication. In: Fields BN, Knipe DM, Howley PM (eds) Fundamental virology. Lippincott-Raven, Philadelphia New York, pp 1 043–1 107

52. Sabin AB (1938) Progression of different nasally instilled viruses along different nervous pathways in the same host. Proc Soc Exp Biol Med 38: 270–275

53. Schmidt J, Klupp BG, Karger A, Mettenleiter TC (1997) Adaptability in herpesviruses: glycoprotein D-independent infectivity of pseudorabies virus. J Virol 71: 17–24

54. Standish A, Enquist LW, Miselis RR, Schwaber JS (1995) Dendritic morphology of cardiac related medullary neurons defined by circuit-specific infection by a recombinant pseudorabies virus expressing beta-galactosidase. J Neurovirol 1: 359–368

55. Tirabassi RS, Townley RA, Eldridge MG, Enquist LW (1997) Characterization of Pseudorabies virus mutants expressing carboxy-terminal truncations of gE: evidence for envelope incorporation, virulence, and neurotropism domains. J Virol 71: 6 455–6 464

56. Tirabassi RS, Enquist LW (1998) The role of envelope protein gE endocytosis in the pseudorabies virus life cycle. J Virol 72: 4 571–4 579

57. Ugolini G (1995) Transneuronal tracing with alpha-herpesviruses: a review of the methodology. In: Kaplitt MG, Loewy AD (eds) Viral vectors. Gene therapy and neuroscience applications. Academic Press, San Diego, pp 293–317

58. Whitley RJ, Schlitt M (1991) Encephalitis caused by herpesviruses including B virus. In: Scheld WM, Whitley RJ, Durack DT (eds) Infections of the central nervous system. Raven Press, New York, pp 41–86

59. Wittmann G (1989) Herpesvirus diseases of cattle, horses and pigs. Kluwer, Boston Dordrecht London

60. Whealy ME, Card JP, Robbins AK, Dubin JR, Rziha H-J, Enquist LW (1993) Specific pseudorabies virus infection of the rat visual system requires both gI and gp63 glycoproteins. J Virol 67: 3 786–3 797

61. Zemanick MC, Strick PL, Dix RD (1991) Direction of transneuronal transport of herpes simplex virus 1 in the primate motor system is strain-dependent. Proc Natl Acad Sci USA 88: 8 048–8 051

62. Zhu Z, Hao Y, Gershon MD, Ambron RT, Gershon AA (1996) Targeting of glycoprotein I(gE) of varicella-zoster virus to the trans-Golgi network by an AYRV sequence and an acidic amino acid-rich patch in the cytosolic domain of the molecule. J Virol 70: 6 563–6 575

63. Tirabassi RS, Enquist LW (1999) Mutation of the YXXL endocytosis motif in the cytoplasmic tail of pseudorabies virus gE. J Virol 73: 2 717–2 728

64. Yang M, Card JP, Tirabassi RS, Miselis RR, Enquist LW (1999) Retrograde, transneuronal spread of pseudorabies virus in defined neuronal circuitry of the rat brain is facilitated by gE mutations that reduce virulence. J Virol 73: 4 350–4 359

65. Brideau A, del Rio T, Wolffe EJ, Enquist LW (1999) Intracellular trafficking and localization of the pseudorabies virus Us9 Type II envelope protein to host and viral membranes. J Virol 73: 4 372–4 384

66. Sparks-Thissen R, Enquist LW (1999) Differential regulation D^k and K^k MHC class I proteins on the cell surface after infection of murine cells by pseudorabies virus. J Virol 73: 5 748–5 756

67. Smith GA, Enquist LW (1999) Construction and transposon mutagenesis in *E. coli* of a full-length infectious clone of pseudorabies virus, an alphaherpesvirus. J Virol (in press)

Author's address: Dr. L. W. Enquist, 314 Schultz Laboratory, Department of Molecular Biology, Princeton University, Princeton, NJ 08544, U.S.A.

Immune modulation by proteins secreted from cells infected by vaccinia virus

G. L. Smith, J. A. Symons, and **A. Alcamí** *

Sir William Dunn School of Pathology, University of Oxford, Oxford, U.K.

Summary. Vaccinia virus comprises the live vaccine that was used for vaccination against smallpox. Following the eradication of smallpox, vaccinia virus was developed as an expression vector that is now used widely in biological research and vaccine development. In recent years vaccinia virus and other poxviruses have been found to express a collection of proteins that block parts of the host response to infection. Some of these proteins are secreted from the infected cell where they bind and neutralise host cytokines, chemokines and interferons (IFN). In this paper three such proteins that bind interleukin (IL)-1β, type I IFNs and CC chemokines are described. The study of these immunomodulatory molecules is enhancing our understanding of virus pathogenesis, yielding fundamental information about the immune system, and providing new molecules that have potential application for the treatment of immunological disorders or infectious diseases.

Introduction

Poxviruses are a group of large DNA viruses that replicate in the cytoplasm [34]. Members of the *Orthopoxvirus* genus have been the most important poxviruses in human medicine. Variola virus caused smallpox, a disease eradicated in 1978 [20], cowpox virus was used by Jenner in 1796 to immunise against smallpox [25], and vaccinia virus is the vaccine that was used this century for smallpox vaccination and whose origin remains an enigma [9]. These viruses are all morphologically indistinguishable and antigenically cross-protective. The genomes are large double stranded (ds) DNA of nearly 200 kb that have been completely sequenced for vaccinia virus strains Copenhagen [21] and MVA [7] and variola strains India-1967 [42] and Bangladesh-1975 [31].

Smallpox was a devastating disease that killed up to 40% of those infected. It struck people of all ages, races and ethnic origins. Like many viruses that cause

*Present address: Division of Virology, Department of Pathology, University of Cambridge, Cambridge, U.K.

a systemic infection, variola spread throughout the body in several phases of replication, and disease symptoms did not develop for more than one week after infection [20]. The failure of the immune system to prevent this severe disease despite the long incubation period suggested that the virus had some strategies to block or escape from the immune system. Studies with other orthopoxviruses, particularly vaccinia virus, have revealed many ways that these viruses do this [47] and it is paradoxical that this knowledge has been acquired only after smallpox was eradicated.

The elimination of smallpox might have been expected to mark the end of interest in vaccinia virus. However, barely had the World Health Organisation certified in 1980 that eradication was complete, than two groups developed techniques to construct recombinant vaccinia viruses expressing foreign genes that had the potential to be used as vaccines against diseases other than smallpox [29, 39]. The proposal to re-use vaccinia virus as a human vaccine was not welcomed by those who recalled the complications that smallpox vaccination had caused [27] and it was generally recognised that before being re-used in human medicine more attenuated vaccinia virus strains were needed. It was the search for virus virulence genes, that when deleted would create safer attenuated vaccines, that led to the discovery of many proteins utilised by the virus for immune evasion. Here three proteins that are secreted from vaccinia virus-infected cells are described. These proteins each bind different soluble host factors that either have direct anti-virus activity or co-ordinate the immune response to infection.

Results

1. Intercepting interleukins (ILs): a soluble virus receptor for IL-1β (vIL-1βR)

The nucleotide sequence of the genome of vaccinia virus strain Western Reserve (WR) revealed an open reading frame (ORF), termed B15R, that contained a signal sequence, 3 domains typical of the immunoglobulin superfamily (IgSF) and sites for addition of N-linked carbohydrate. However, there was no transmembrane anchor sequence and cytoplasmic domain and so the protein was predicted to be a secreted glycoprotein [44]. Of particular interest was the 25% amino acid similarity of B15R with the extracellular ligand binding domain of the type I interleukin-1 receptor (IL-1R), and hence it was predicted that this molecule might function as a soluble IL-1R [44]. Shortly afterwards, the type II IL-1R was cloned and this was found to be more closely related to the B15R protein than B15R was to the type I IL-1R [32].

Two independent research groups then demonstrated that the B15R protein functioned as an IL-1R [2, 48]. The following description relates mostly to work from our laboratory. The B15R gene was shown to encode a secreted glycoprotein of 50–60 kDa that is expressed late during the infectious cycle [2]. To test if the B15R protein functioned as an IL-1R, the protein was overexpressed from vaccinia virus (vB15R) and recombinant baculovirus (AcB15R), and a vaccinia virus WR deletion mutant lacking the gene (vΔB15R) was constructed. Supernatants from

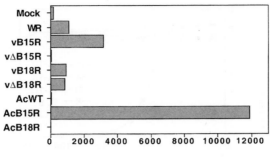

Fig. 1. Vaccinia virus WR B15R protein functions as a soluble receptor for IL-1β. Human TK⁻143 cells were either mock-infected or infected with vaccinia viruses WR, vB15R, vΔB15R, vB18R (a strain of vaccinia virus WR overexpressing vaccinia virus WR gene B18R) or vΔB18R (a strain of vaccinia virus WR from which the B18R gene had been deleted) at 10 pfu/cell and the supernatants were harvested at 24 h post infection. Alternatively, Spodoptera frugiperda (Sf) 21 cells were infected with Autographa californica nuclear polyhedrosis virus (AcNPV), or recombinant baculoviruses expressing the B15R (AcB15R) or B18R (AcB18R) genes from vaccinia virus strain WR at 5 pfu/cell and the supernatants were harvested 3 days post infection. Supernatants were centrifuged to remove cellular debris and the clarified supernatants were used in binding assays with 180 pM of human ¹²⁵I-IL-1β. Complexes formed between the virus IL-1βR and ¹²⁵I-IL-1β were precipitated with polyethylene glycol, the precipitate captured by filtration and the radioactivity counted as described previously [2]. Reproduced with permission from Cell Press, from Alcamí A, Smith GL (1992) A soluble receptor for interleukin-1β encoded by vaccinia virus: a novel mechanism of virus modulation of the host response to infection. Cell 71: 153–167

cells infected with wild type (WT) vaccinia, vB15R and AcB15R each contained an activity that bound human ¹²⁵I-IL-1β but this activity was absent from mock-infected cells and from cells infected with vΔB15R [2] (Fig. 1). These data showed that B15R encoded a virus IL-1β receptor (vIL-1βR) and that this was the only vaccinia virus gene that did so.

IL-1 comes in three forms: IL-1α, IL-1β and the IL-1 receptor antagonist (ra) protein that negatively regulates IL-1 activity by binding to IL-1 receptors without inducing signal transduction [19]. To determine which types of IL-1 were bound by B15R, human ¹²⁵I-IL-1β was incubated with B15R in the presence of increasing concentrations of unlabelled human IL-1α, IL-1β or IL-1ra. Only IL-1β was able to compete with ¹²⁵I-IL-1β for binding to vIL-1βR, indicating that vIL-1βR was specific for IL-1β (Fig. 2) [2]. This was true for both human and murine IL-1 [4].

Scatchard analysis showed that approximately 10^5 copies of the B15R protein were released from each infected cell in a 24 h period and that the molecule had a high affinity for IL-1β (Kd=226±38 pM) [2]. This high affinity suggested that vIL-1βR would compete effectively with cellular IL-1Rs for IL-1β, a prediction confirmed in competition experiments with EL4 cells bearing IL-1Rs. In the absence of competitor, ¹²⁵I-IL-1β bound to these cells, but this binding was prevented by vIL-1βR or unlabelled IL-1β [2]. This established that

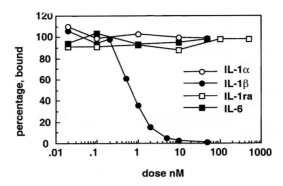

Fig. 2. The vaccinia virus IL-1βR binds only IL-1β. Medium derived from 4×10^4 TK$^-$143 cells that had been infected with vaccinia virus WR as described in Fig. 1 was incubated with 100 pM of human ^{125}I-IL-1β in the presence of unlabelled IL-1α (open circles), IL-1β (closed circles), IL-1ra (open squares) or IL-6 (closed squares). The radioactivity bound to vIL-1βR was precipitated with polyethylene glycol and counted as described in Fig. 1. Reproduced with permission from Cell Press, from Alcamí A, Smith GL (1992) A soluble receptor for interleukin-1β encoded by vaccinia virus: a novel mechanism of virus modulation of the host response to infection. Cell 71: 153–167

vIL-1βR could inhibit IL-1β binding to cells. In a parallel study vIL-1βR was shown to block the ability of IL-1β to induce synthesis of IL-2 [48].

Role of the vIL-1βR in virus pathogenesis

To determine the role of the vIL-1βR in virus pathogenesis, WT and deletion mutant (vΔB15R) viruses were inoculated intranasally into BALB/c mice. Surprisingly, infection by this route with the deletion mutant virus gave rise to a more severe infection than with the WT virus [2]. In contrast, if similar viruses were inoculated by intracranial injection, the deletion mutant showed a lower virulence than WT [48]. In the intranasal model, animals infected with vΔB15R showed enhanced weight loss and signs of illness (pilo-erection, arched back and reduced mobility), and although the overall number of mortalities was unchanged compared to infection with WT virus, the mortalities induced by the deletion mutant occurred sooner [2].

To understand how the deletion mutant was able to induce a more severe illness, the function of IL-1 was considered. This pro-inflammatory cytokine can induce local mediators of inflammation, such as IL-2, that might help to restrict virus replication, but high levels of IL-1 can induce systemic effects such as fever, weight loss, headache and shock [19]. Infection with either WT or deletion mutant virus led to a serious illness that was likely to induce IL-1β production. However, after infection with WT virus, vIL-1βR was released from virus-infected cells at sufficient levels to be detected systemically [4]. Consequently, vIL-1βR would bind available IL-1β and reduce the levels of this cytokine. In contrast, after infection by vΔB15R, the levels of IL-1β would not be reduced by vIL-1βR and so might be elevated and contribute to the disease. To examine this further, the

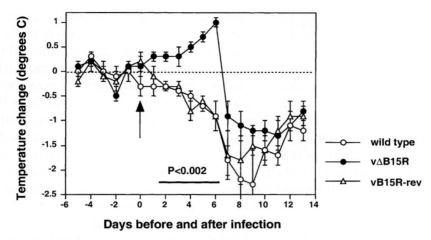

Fig. 3. The vIL-1βR prevents fever in infected mice. Groups of 10 female BALB/c mice were infected intranasally with 2.5×10^3 pfu of either plaque purified WT vaccinia virus WR (wild type), or vΔB15R or vB15R-rev. The mean change in rectal temperature (±SEM) and the basal temperature (broken line) are shown. The horizontal bar indicates those days on which the difference between vΔB15R and both WT and vB15R-rev were statistically significant when analysed by the Student's *t* test and the mean P value is shown. The arrow indicates the day of infection. Reproduced with permission from the National Academy of Sciences, USA from Alcamí A, Smith GL (1996) A mechanism for the inhibition of fever by a virus. Proc Natl Acad Sci USA 93: 11 029–11 034 ©1996 National Academy of Sciences, U.S.A.

body temperature of infected animals was measured, because IL-1β was one of several cytokines reported to affect fever.

vIL-1βR prevents fever in infected animals

Mice that were infected intranasally with a plaque purified, WT, vaccinia virus WR developed a slight hypothermia for several days after infection, despite suffering a severe infection that resulted in substantial loss of body weight. In contrast, animals infected with vΔB15R developed a fever that was sustained for 5 days (Fig. 3). Infection with a revertant virus, vB15R-rev, in which the B15R gene was re-inserted into the deletion mutant, induced a temperature profile that was indistinguishable from that induced by WT virus. This confirmed that it was the loss of the B15R protein, and not some other mutation elsewhere in the large virus genome, that caused the temperature difference [4]. The demonstration that this virus protein controlled the temperature of the infected animals was novel for virus pathogenesis but also had implications for the physiological role of cytokines: because vIL-1βR bound only IL-1β and not other endogenous cytokines such IL-1α, IL-6, tumour necrosis factor (TNF), interferon (IFN)-α/β and IFN-γ, the result demonstrated that it was IL-1β and not these other putative pyrogens that was controlling fever [4]. A similar conclusion was reached independently using transgenic mice lacking IL-1β that were unable to produce fever in response to pyrogenic stimuli [26].

Two other observations were consistent with the B15R protein controlling the body temperature of infected animals. First, during the smallpox eradication campaign many different strains of vaccinia virus were used and these differed in their virulence for humans [20]. In particular, the Copenhagen and Tashkent strains had been more virulent, as judged by the frequency of post-vaccinial complications such as encephalitis [20]. Other strains such as Lister and Wyeth were safer and produced lower rates of this and other complications [20]. The expression of a vIL-1βR by these strains was examined and, in accord with the result observed with the WR strain in mice, it was found that the more virulent strains Copenhagen and Tashkent did not express vIL-1βR but did induce fever in infected mice [4]. This might have been coincidence as these viruses probably contain several differences in addition to whether or not they express vIL-1βR. Therefore, the Copenhagen strain was studied in more detail.

The complete nucleotide sequence of the Copenhagen strain had revealed that the vIL-1βR gene (B16R in strain Copenhagen) contained a frameshift mutation near the 5' end of the ORF [21]. The gene was therefore repaired and a recombinant Copenhagen virus that expressed vIL-1βR was constructed. Like the WT WR strain, but unlike the parental Copenhagen virus, this virus was unable to induce fever in infected mice despite inducing a severe infection.

To be absolutely certain that it was the binding of only IL-1β by vIL-1βR that was causing the change in body temperature during infection, mice were infected with the WT Copenhagen virus and then injected with a monoclonal antibody that neutralises IL-1β. The fever that mice developed after infection with WT Copenhagen was prevented by the injection of the IL-1β specific antibody [4]. This result taken together with the known specificity of the vIL-1βR for only IL-1β, confirmed that in this model of infection IL-1β was the principal endogenous pyrogen.

The vIL-1βR gene is non-functional in variola viruses

The DNA sequence of variola virus is related very closely to vaccinia virus over the central region of the genome, but differs significantly near the genomic termini where genes are found that are unique to either virus [1, 31, 42]. A surprising difference between these viruses was that several genes in variola virus that were closely related to vaccinia virus counterparts (> 90% nucleotide identity) were disrupted and presumed non-functional. This included the B15R gene of vaccinia virus that encoded the vIL-1βR. In variola viruses India-1967 [42] and Bangladesh-1975 [31] this gene was disrupted in multiple places and is non-functional. The disease smallpox was characterised by a severe illness with high fever, and the lack of expression of vIL-1βR by variola virus was consistent with the fever induced by infection with this poxvirus.

2. Interfering with interferon

Vaccinia virus produces several proteins that inhibit the action of IFN either within the infected cell or by preventing IFN binding to its natural receptor and inducing expression of IFN-responsive genes, for review see [46]. The E3L and K3L proteins function inside the infected cell and block the action of IFN-induced anti-virus proteins. K3L is a 10.5 kDa protein with amino acid similarity to the eukaryotic translation initiation factor 2α (eIF2α) [10]. Phosphorylation of eIF2α by the IFN-induced and dsRNA-activated protein kinase PKR causes an inhibition of protein synthesis. K3L serves as a substrate for PKR so that the phosphorylation of eIF2α is prevented and protein synthesis can continue. E3L is a 25 kDa dsRNA-binding protein [15]. Both PKR and the IFN-induced enzyme 2′5′-oligoadenylate synthetase are activated by binding dsRNA that commonly is produced during virus infection. The E3L protein binds the dsRNA so that neither enzyme is activated and protein synthesis can continue.

In addition to these intracellular proteins that protect virus-infected cells against the anti-virus effects of IFN, most orthopoxviruses also secrete proteins that bind type I IFNs (IFN-α/β) and type II IFN (IFN-γ) in solution. The first poxvirus IFN-binding protein was discovered in myxoma virus [54]. This protein shares amino acid similarity with the extracellular domain of the cellular type II IFN receptor (IFN-γR) and with a similar sized protein from vaccinia virus [24, 54]. The protein from vaccinia virus and other orthopoxviruses was found to bind and inhibit IFN-γ from a wide variety of species [3, 36, 37].

A soluble receptor for type I IFN (vIFN-α/βR)

Independently, two groups found an inhibitor of type I IFNs in the supernatant of vaccinia virus-infected cells [17, 49]. In this laboratory, the IFN inhibitor was identified by screening supernatants from infected cells for activity that could inhibit the anti-virus activity of IFN. In the absence of IFN, the infection of HeLa cells by cocal virus, a rhabdovirus related to vesicular stomatitis virus, led to the destruction of the cells within 2 days. But, if the cells were pre-treated with increasing doses of IFN, the cells were protected from destruction by the subsequent virus infection. However, the anti-virus action of IFN was overcome if the IFN was administered together with the supernatant from vaccinia virus WR-infected, but not mock-infected, cells. Thus the supernatant of the infected cells contained an IFN inhibitor. This inhibitor was widely distributed in orthopoxviruses [49] (Fig. 4) but vaccinia virus strains Lister and MVA [13] did not express the activity and Wyeth expressed an IFN inhibitor with a much lower affinity for type I IFN [49]. Notably these viruses were among the safer strains of vaccinia virus used for smallpox vaccination [20].

Mapping the gene encoding the type I IFN inhibitor

The vaccinia virus Copenhagen genome had been sequenced [21], but we found no protein with convincing amino acid similarity to the cellular type I IFN receptors. Consequently, the gene was mapped by molecular genetics. To do this we took advantage of the fact that most of the virus genes that affected host range or virulence are located towards the genomic termini, and that several mutant viruses were available that had large deletions in these regions [12, 35].

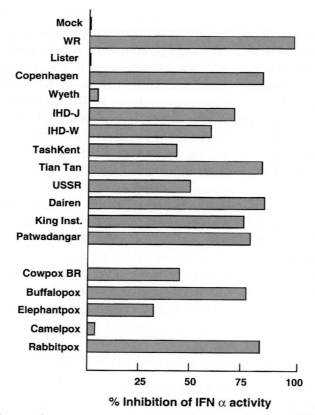

Fig. 4. Most orthopoxviruses express a soluble inhibitor of human IFNα. TK⁻143 cells were infected with the indicated strains of vaccinia virus or other orthopoxviruses and the supernatant from 10^3 cells was tested for its ability to inhibit 5 units of human natural IFNα. The test culture supernatant and IFNα were mixed and incubated with HeLa cells for 18 h. These cells were then infected with 100 pfu of cocal virus and 48 h later the number of plaques was counted. Data are expressed as the percentage of IFNα activity that has been inhibited by the virus supernatant, i.e. the number of plaques formed after treatment of cells with IFN and test supernatant divided by the number of plaques formed by cocal virus in the absence of IFN. Reproduced with permission from Cell Press, from Symons JA, Alcamí A, Smith GL (1995) Vaccinia virus encodes a type I interferon receptor of novel structure and broad species specificity. Cell 81: 551–560

Supernatants from cells infected with these different mutants were screened for the type I IFN inhibitor and this led to the identification of a virus (vSSK2) that lacked the activity [49]. Importantly, this virus had a large deletion near the right end of the genome but differed from another virus (vGS100) that expressed the type I IFN inhibitor, by the presence of only nine genes. Only two of these genes (B15R and B18R) were predicted to encode a secreted glycoprotein, and one of these (B15R) had already been characterised as the vIL-1βR [2]. B18R was shown to encode the anti-IFN activity by using a virus lacking the B18R gene (vΔB18R) and a recombinant baculovirus that expressed this gene (AcB18R). Supernatants from vaccinia virus WR-infected cells, but not mock-infected or

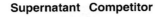

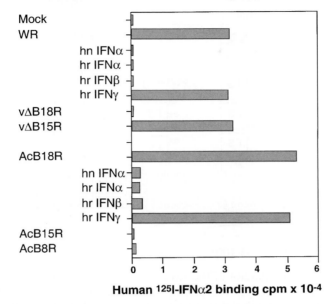

Human ¹²⁵I-IFNα2 binding cpm x 10⁻⁴

Fig. 5. Vaccinia virus WR gene B18R encodes the soluble IFNα inhibitor. Supernatants from 10^5 TK⁻ 143 cells that had been mock-infected or infected with vaccinia virus WR, v△B15R or v△B18R, or supernatants from Sf cells that had been infected with baculoviruses AcB15R, AcB18R or AcB8R (a recombinant baculovirus expressing the vaccinia virus WR B8R gene [3]) were incubated with 50 nM of human ¹²⁵I-IFNα2 in the presence or absence of a 100 fold excess of unlabelled human IFNα, IFNβ, or IFNγ. The vIFN-α/βR-¹²⁵I-IFNα complex was precipitated with polyethylene glycol, captured by filtration and the radioactivity counted. Reproduced with permission from Cell Press, from Symons JA, Alcamí A, Smith GL (1995) Vaccinia virus encodes a type I interferon receptor of novel structure and broad species specificity. Cell 81: 551–560

v△B18R-infected cells, contained a protein that bound human ¹²⁵I-IFNα2 (Fig. 5). The B18R protein bound only type I IFNs, as shown by the ability of an excess of unlabelled type I IFNs but not type II IFN to prevent the formation of the ¹²⁵I-IFNα2-B18R complex. Similarly, AcB18R, but not AcB15R, expressed a type I IFN-binding protein. Scatchard analysis demonstrated that the B18R protein had a high affinity constant for human IFNα2 (Kd=174±15 pM) [49]. Hereafter the B18R protein is referred to as vIFN-α/βR.

Like vIL-1βR, vIFN-α/βR has a signal peptide, three Ig domains, sites for addition of N-linked carbohydrate, but no transmembrane anchor sequence or cytoplasmic domain [44]. It was surprising that vIFN-α/βR was an IgSF because the cellular type I IFN receptors that had been cloned and sequenced [38, 55] were members of a different protein superfamily (the class II cytokine receptor family) and contained 2 or 4 copies of fibronectin type III repeats (Fig. 6). Yet the virus and cellular proteins bound the same ligand with comparable affinities. Possibly vIFN-α/βR represents a virus version of a cellular protein that has yet to be cloned, or alternatively, the virus and cellular proteins may represent an example

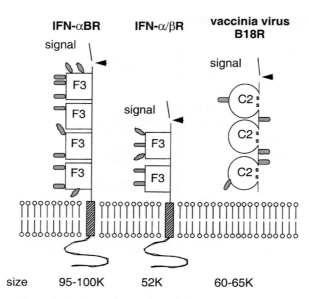

Fig. 6. Schematic representation of the structure of the vaccinia virus vIFN-α/βR (B18R) and the two cellular type I IFN receptors [38, 55]. The cellular type I IFN receptors are membrane bound glycoproteins that contain either 2 or 4 copies of the fibronectin type III repeat and are members of the class II cytokine receptor superfamily. In contrast, the vIFN-α/βR is a soluble glycoprotein that is secreted from infected cells and is a member of the IgSF with three Ig constant region type 2 domains. Note the vIFN-α/βR is also the poxvirus S antigen [51] and binds to the cell surface via an unidentified cellular molecule. Arrowheads indicate sites of proteolytic cleavage to remove the signal peptide, shaded areas indicate potential sites for attachment of carbohydrate via asparagine-residues and hatched areas indicate transmembrane anchor sequences. The size of the polypeptides in Daltons is indicated at the bottom

of convergent evolution. Other investigators reported limited amino acid similarity between regions of the cellular type I IFN receptor and vIFN-α/βR despite these proteins having different types of repeated domains [17]. A structural study of the virus protein complexed with type I IFN is needed to determine exactly how vIFN-α/βR interacts with its ligands, but there is already an indication that vIFN-α/βR and cellular IFN receptors bind type I IFNs differently. Only monoclonal antibodies directed against the N-terminal region of human type I IFNs inhibited interaction of these IFNs with the cellular IFN receptor, whereas antibodies to both N- and C-terminal regions of IFN were unable to recognise IFN after it was bound by vIFN-α/βR [28].

The vIFN-α/βR is also present on the cell surface

Since the 1960s orthopoxviruses have been known to express an antigen, called the S antigen, that is present on the surface of infected cells [6, 50, 52, 53]. The finding that vaccinia virus strain Lister did not express this antigen enabled the gene encoding the S antigen to be identified as B18R by molecular genetics [51]. The presence of vIFN-α/βR on the cell surface, despite the lack of a C-terminal hydrophobic transmembrane domain, led to the proposal that vIFN-α/βR was attached to cells by adopting a type II membrane topology [33]. However, an alternative view is supported by the observation that the soluble vIFN-α/βR collected from the supernatant of AcB18R-infected insect cells, could bind to uninfected mammalian cells via an unidentified cellular molecule [17] (Alcamí, Symons and Smith, unpubl. data).

Once IFN binds to the appropriate cell surface receptor it induces signal transduction and phosphorylation of JAK protein kinases and STAT transcription factors. vIFN-α/βR bound to the cell surface was shown to prevent such signal transduction and phosphorylation of proteins in response to addition of IFN [17].

vIFN-α/βR binds IFNs from a broad range of species

IFNs are usually highly species specific and so it was possible that vIFN-α/βR might bind to type I IFN from only a single species. If so, this would have been very interesting in view of the unknown origin and natural host of vaccinia virus [9], and might indicate where the virus, or more particularly the gene, came from. Therefore, the specificity of vIFN-α/βR for type I IFNs from different species was investigated. Human ^{125}I-IFNα2 was mixed with vIFN-α/βR in the presence of increasing concentrations of unlabelled type I IFNs from different species and the receptor-ligand complex was precipitated, captured by filtration and the radioactivity was counted. This demonstrated that human and rabbit type I IFNs competed with ^{125}I-IFNα2 for binding to vIFN-α/βR with similar efficiency, indicating they had similar affinities for vIFN-α/βR. Rat and bovine type I IFNs were 5- to 10-fold less effective competitors, but still much better than mouse type I IFNs that were 2 to 3 orders of magnitude less effective competitors than rabbit IFN. The same conclusions were reached using biological assays [49]. Thus, vIFN-α/βR had a very broad species specificity that was without precedent in the IFN system. Sadly, this result did not indicate any specific species as the likely host for vaccinia virus, although a mouse seemed an unlikely host because of the low affinity for mouse type I IFNs. However, the broad species specificity suggested that vIFNα/βR would have been a very useful protein during virus evolution, particularly if the virus replicated in multiple species.

The role of vIFN-α/βR in virus virulence

vIFN-α/βR is important in vivo and contributes to virus virulence. The majority of female BALB/c mice infected intranasally with 10^4 or 10^5 plaque forming units (pfu) of either a plaque purified vaccinia virus WR or a revertant virus (vB18R-rev) in which the B18R gene had been re-inserted into the vΔB18R deletion mutant, died with 14 days after infection. However, all the animals infected with these doses of vΔB18R survived the infection [49]. Moreover, the weight loss in vΔB18R-infected animals was much less than in control-virus infected animals. Similarly, the titres of virus present in lungs and brain after infection with the deletion mutant were much lower than in WT or vB18R-rev-infected animals [49]. This virus attenuation in mice was observed despite the fact that vIFNα/βR bound mouse type I IFNs poorly compared to IFNs from other species such as man. It follows that the vΔB18R virus would probably be more attenuated compared to WT virus in man. In this regard, it is noteworthy that the three vaccinia virus strains, Lister, modified virus Ankara and Wyeth that produced lower frequencies of vaccine related complications in man, either do not express vIFNα/βR or, in the case of Wyeth, produce a protein with an 85-fold reduced affinity for human IFNα2 [13, 49].

3. A soluble chemokine binding protein from vaccinia virus

Chemokines are small soluble proteins that are chemoattractant for various types of leukocytes that bear appropriate chemokine receptors (CKRs) on their surface and are classified into several sub-groups based upon the number and arrangement of conserved cysteine residues near their amino terminus [8]. CC chemokines are exemplified by RANTES, eotaxin, and macrophage inflammatory protein (MIP)-1α, and have two consecutive C residues. CXC chemokines, exemplified by IL-8 and GRO-α, contain a single amino acid between the conserved cysteines. Lymphotactin is the sole member of the C chemokine sub-group and has only a single conserved cysteine. Fractalkine is the sole member of a CX$_3$C chemokine subgroup and is unusual in being membrane-associated and in having several other domains in addition to the chemokine domain.

Chemokine receptors (CKRs) are 7 transmembrane G-protein coupled proteins that are highly hydrophobic [41]. Consequently, soluble versions of these molecules that bind chemokines have not been described. CKRs have received great interest since 1996 when it was recognised that these proteins are used by human immunodeficiency virus (HIV) as a co-receptor on CD4 positive cells [11]. In addition, other pathogens such as the malarial parasite *Plasmodium vivax*, also use these molecules to initiate infection [23].

The sequence of the vaccinia virus Copenhagen genome revealed that there were no proteins related to known CKRs [21]. Nonetheless, three research groups identified a soluble chemokine binding protein (CKBP) that is released from cells infected with some strains of vaccinia virus and other orthopoxviruses [5, 22, 43]. In our group, the virus CKBP (vCKBP) was identified by incubating supernatants from infected cells with ^{125}I-RANTES, chemically crosslinking any complex that was formed and then subjecting the sample to polyacrylamide gel electrophoresis followed by autoradiography. If a vCKBP was present in the culture supernatant, the electrophoretic mobility of the chemokine would be retarded. This analysis demonstrated that the supernatants of cells infected by the majority of orthopoxviruses contained a vCKBP that reacted with the human CC chemokine RANTES (Fig. 7). However, not all vaccinia virus strains expresses vCKBP and notably the commonly used strains WR and Copenhagen lacked the activity [5].

The gene encoding the vCKBP was identified by matching the pattern of vaccinia virus strains that expressed vCKBP with the pattern of viruses that

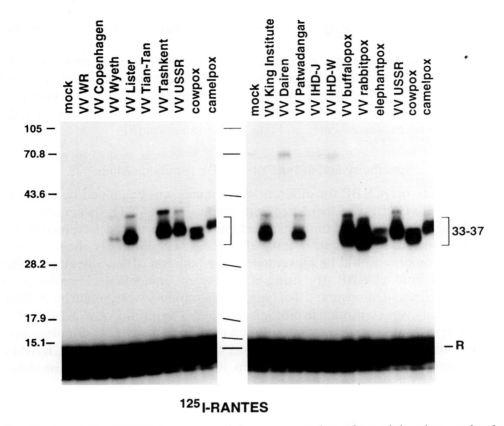

Fig. 7. A soluble CKBP is expressed by some strains of vaccinia virus and other orthopoxviruses. Media from TK⁻143 cells that had been mock-infected or infected with the indicated orthopoxviruses were prepared as described in Fig. 1 and incubated with ¹²⁵I-RANTES. Samples were treated with the chemical cross-linker 1-ethyl-3-(3-dimethylaminopropyl)-carbodiimide (*EDC*) and then subjected to SDS-PAGE analysis and autoradiography. The position of RANTES (*R*) and the RANTES-vCKBP complexes (square bracket), and the sizes of proteins in kiloDaltons are indicated. Reproduced with permission from Alcamí A, Symons JA, Collins P, Williams TJ, Smith GL (1998) Blockade of chemokine activity by a novel soluble chemokine binding protein from vaccinia virus. J Immunol 160: 624–633. ©1998 The American Association of Immunologists

expressed a 35 kDa soluble protein of unknown function [40]. The gene encoding the 35 kDa protein resides within the inverted terminal repeat of vaccinia virus strain Lister and this gene was shown to encode vCKBP in several ways: i) the WT vaccinia virus Lister strain expressed vCKBP, but a mutant virus engineered so that the gene encoding this protein was disrupted did not; ii) the 35 kDa protein expressed by recombinant baculovirus (Ac35K) bound RANTES; iii) the 35 kDa protein expressed as a chimera fused to the Fc portion of human immunoglobulin G produced a fusion protein that bound RANTES; iv) antibody directed against the 35 kDa protein inhibited the formation of a ¹²⁵I-RANTES-35 kDa protein complex. The specificity of the interaction of vCKBP with CC chemokines was confirmed by inhibiting the binding of these recombinant proteins to ^{125}I-RANTES by an excess of unlabelled CC chemokines but not IFN-γ [5].

The promoter for the gene encoding vCKBP had been used extensively in the construction of recombinant vaccinia viruses long before the function of the intact gene was known and had been called the 7.5K promoter [29, 45]. This promoter was active early and late during infection [16, 30]. Consistent with this, vCKBP was found to be expressed early and late during infection [5].

Specificity and affinity of vCKBP for chemokines

The specificity of vCKBP for different chemokines was examined by competition assays. If vCKBP was incubated with human ^{125}I-RANTES a complex was formed and the formation of this radiolabelled complex could be prevented by the addition of increasing amounts of unlabelled RANTES. Similarly, the formation of the ^{125}I-complex would be prevented by the addition of any chemokine that was able to bind vCKBP. By using a variety of unlabelled chemokines as competitors, it was demonstrated that only CC chemokines were bound by vCKBP. This specificity was confirmed in biological assays. The affinity of vCKBP for CC chemokines was determined by scintillation proximity assay [14]. Fluoromicrospheres covered with protein A (Amersham) were incubated with the vCKBP-Fc fusion protein and the complex was then incubated with ^{125}I-MIP-1α. Only

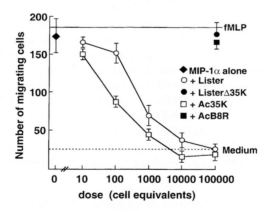

Fig. 8. Inhibition of chemotaxis by vCKBP. U937 cells (2×10^4) were placed in the upper chamber of a 24 well plate containing a 10 mm tissue culture insert. MIP-1α (50 ng/ml) with or without the indicated test reagents was placed in the lower chamber separated by a polycarbonate filter with a pore size of 8 μm. After incubation for 2 h at 37 °C the filter was washed, stained with 4,6,-diamidino-2-phenylindole (DAPI) and the number of migrated cells observed in 5 fields (±SEM) in a fluorescent microscopy was counted. The line indicates the number of cells that had migrated in the absence of competitor or in the presence of fMLP. Reproduced with permission from Alcamí A, Symons JA, Collins P, Williams TJ, Smith GL (1998) Blockade of chemokine activity by a novel soluble chemokine binding protein from vaccinia virus. J Immunol 160: 624–633. ©1998 The American Association of Immunologists

radioactivity bound to the beads was close enough to excite the scintillant within the fluoromicrospheres and cause an emission, so that separation of bound and unbound ^{125}I-MIP-1α was not necessary to enable detection of bound ligand. This enabled a saturation curve to be obtained using increasing amounts of with ^{125}I-MIP-1α and Scatchard analysis of this curve led to determination of an affinity constant. vCKBP had a very high affinity for MIP-1α (Kd=104±4 pM), equivalent to or up to 100 fold higher than the affinity of CC chemokines for cellular CKRs. The use of different cold competitors enabled the affinity of vCKBP for eotaxin, RANTES, and MCP to be determined as Kd=1.4±0.6 nM, 7.2±0.8 nM and 15.1±0.6 nM, respectively [5].

Such affinities suggested that vCKBP would compete effectively with cellular CKRs for binding of CC chemokines. This was shown by incubating MIP-1α with U937 cells (a human monocytic cell line) in the presence of increasing concentrations of vCKBP derived from the supernatants of vaccinia virus strain Lister-infected human cells or Ac35K-infected insect cells. In the absence of the competitor, MIP-1α bound to U937 cells, but this interaction was inhibited by vCKBP. The specificity of vCKBP for CC chemokines was confirmed by its inability to inhibit the binding of a CXC chemokine GRO-α to the same cells [5].

Inhibition of calcium flux and chemotaxis in vitro

Binding of a chemokine to its receptor triggers a transient increase in intracellular calcium concentration. Consequently, because vCKBP prevented the binding of CC chemokines to CKRs, it was predicted that vCKBP would block this calcium flux. This was demonstrated using human eosinophils and eotaxin. The calcium flux induced in eosinophils by addition of eotaxin was inhibited by addition of vCKBP. In contrast, the calcium flux induced by the binding of CXC chemokines IL-8 or GRO-α on human neutrophils was not inhibited by vCKBP. This confirmed the specificity of vCKBP for CC chemokines in a biological assay.

The inhibition of calcium mobilisation by vCKBP would be expected to prevent the migration of a leukocyte bearing an appropriate CKR towards a chemokine. This was tested by measuring the migration of U937 cells across a filter towards a solution of MIP-1α (Fig. 8). The migration of these cells was inhibited if MIP-1α mixed with vCKBP that had been produced from Ac35K-infected cells or from vaccinia virus strain Lister-infected cells. In contrast, no inhibition was observed with control proteins or with the supernatants from cells infected with the Lister virus lacking the gene encoding vCKBP.

Inhibition of eosinophilia in vivo

vCKBP was shown to inhibit the action of the CC chemokine eotaxin in an in vivo model. If guinea pigs are injected with IL-5 and then eotaxin is injected locally into the skin, the mobilised eosinophils are recruited to the site of injection of eotaxin [18]. This accumulation of eosinophils was prevented if the eotaxin was co-injected with vCKBP but not control proteins. Eosinophil infiltration is a feature of allergic inflammatory reactions such as asthma, and consequently such conditions might be prevented or treated by vCKBP or its derivatives.

Conclusions

Poxviruses are a group of large DNA viruses that have played important roles in virology, immunology and medicine. Smallpox was a devastating human disease that was caused by variola virus and was controlled and eventually eradicated by vaccination. Cowpox virus and more recently vaccinia virus represent the first human vaccine and the only one so far to have resulted in the eradication of a human disease. Vaccinia virus was the first animal virus to be physically purified, accurately titrated and chemically analysed. It was with vaccinia virus that virus polymerases were discovered in 1967, a finding that prompted the search for similar enzymes in other viruses and culminated in the discovery of reverse transcriptase a few years later. More than 30 years later new enzymes are still being discovering in this virus. In the early 1980s vaccinia virus pioneered the concept of using genetically engineered live recombinant viruses as vaccines, a principle now applied to many other viruses and micro-organisms. More recently, vaccinia virus and other poxviruses have been found to express a wide range of proteins that inhibit components of the host response to infection. This is illustrated here with three examples of vaccinia virus proteins that are released from infected cells and which bind to IL-1β, type I IFNs and CC chemokines. These molecules are only examples of a wide range of proteins that are secreted from poxviruses infected cells. Others inhibit inflammation, complement, or type II IFN or act as growth factors to stimulate the growth of surrounding cells. Yet other secreted proteins are known to affect virus virulence but the mechanism by which these proteins function is unknown. The study of these virus proteins is an exciting area of research and may lead to the discovery of new proteins that function in the immune system and which are as yet unknown.

Acknowledgement

The work in this laboratory was supported by The Wellcome Trust, the Medical Research Council and the EC.

References

1. Aguado B, Selmes IP, Smith GL (1992) Nucleotide sequence of 21.8 kbp of variola major virus strain Harvey and comparison with vaccinia virus. J Gen Virol 73: 2 887–2 902
2. Alcamí A, Smith GL (1992) A soluble receptor for interleukin-IL-1β encoded by vaccinia virus: a novel mechanism of virus modulation of the host response to infection. Cell 71: 153–167
3. Alcamí A, Smith GL (1995) Vaccinia, cowpox, and camelpox viruses encode soluble gamma interferon receptors with novel broad species specificity. J Virol 69: 4 633–4 639
4. Alcamí A, Smith GL (1996) A mechanism for the inhibition of fever by a virus. Proc Natl Acad Sci USA 93: 11 029–11 034
5. Alcamí A, Symons JA, Collins PD, Williams TJ, Smith GL (1998) Blockade of chemokine activity by a soluble chemokine binding protein from vaccinia virus. J Immunol 160: 624–633

6. Amano H, Ueda Y, Tagaya I (1979) Orthopoxvirus strains defective in surface antigen induction. J Gen Virol 44: 265–269
7. Antoine G, Scheiflinger F, Dorner F, Falkner FG (1998) The complete genomic sequence of the modified vaccinia Ankara strain: comparison with other orthopoxviruses. Virology 244: 365–396
8. Baggiolini M, Dewald B, Moser B (1997) Human chemokines: an update. Annu Rev Immunol 15: 675–705
9. Baxby D (1981) Jenner's smallpox vaccine. The riddle of the origin of vaccinia virus. Heinemann, London
10. Beattie E, Tartaglia J, Paoletti E (1991) Vaccinia-virus encoded eIF-2α homolog abrogates the antiviral effect of interferon. Virology 183: 419–422
11. Berger EA (1997) HIV entry and tropism: the chemokine receptor connection. AIDS 11: S3–16
12. Blake NW, Kettle S, Law KM, Gould K, Bastin J, Townsend AR, Smith GL (1995) Vaccinia virus serpins B13R and B22R do not inhibit antigen presentation to class I-restricted cytotoxic T lymphocytes. J Gen Virol 76: 2 393–2 398
13. Blanchard TJ, Alcamí A, Andrea P, Smith GL (1998) Modified vaccinia virus Ankara undergoes limited replication in human cells and lacks several immunomodulatory proteins: implications for use as a human vaccine. J Gen Virol 79: 1 159–1 167
14. Bosworth N, Towers P (1989) Scintillation proximity assay. Nature 341: 167–168
15. Chang H-W, Watson JC, Jacobs BL (1992) The E3L gene of vaccinia virus encodes an inhibitor of the interferon-induced, double-stranded RNA-dependent protein kinase. Proc Natl Acad Sci USA 89: 4 825–4 829
16. Cochran MA, Puckett C, Moss B (1985) In vitro mutagenesis of the promoter region for a vaccinia virus gene: evidence for tandem early and late regulatory signals. J Virol 54: 30–37
17. Colamonici OR, Domanski P, Sweitzer SM, Larner A, Buller RML (1995) Vaccinia virus B18R gene encodes a type I interferon-binding protein that blocks interferon alpha transmembrane signaling. J Biol Chem 270: 15 974–15 978
18. Collins PD, Marleau S, Griffiths-Johnson DA, Jose PJ, Williams TJ (1995) Cooperation between interleukin-5 and the chemokine eotaxin to induce eosinophil accumulation in vivo. J Exp Med 182: 1 169–1 174
19. Dinarello CA (1996) Biological basis for interleukin-1 in disease. Blood 87: 2 095–2 147
20. Fenner F, Anderson DA, Arita I, Jezek Z, Ladnyi ID (1988) Smallpox and its eradication. World Health Organisation, Geneva
21. Goebel SJ, Johnson GP, Perkus ME, Davis SW, Winslow JP, Paoletti E (1990) The complete DNA sequence of vaccinia virus. Virology 179: 247–266
22. Graham KA, Lalani AS, Macen JL, Ness TL, Barry M, Liu L, Lucas A, Clark-Lewis I, Moyer RW, McFadden G (1997) The T1/35 kDa family of poxvirus-secreted proteins bind chemokines and modulate leukocyte influx into virus-infected tissues. Virology 229: 12–24
23. Horuk R, Chitnis CE, Darbonne WC, Colby TJ, Rybicki A, Hadley TJ, Miller LH (1993) A receptor for the malarial parasite Plasmodium vivax: the erythrocyte chemokine receptor. Science 261: 1 182–1 184
24. Howard ST, Chan YS, Smith GL (1991) Vaccinia virus homologues of the Shope fibroma virus inverted terminal repeat proteins and a discontinuous ORF related to the tumor necrosis factor receptor family. Virology 180: 633–647
25. Jenner E (1798) An enquiry into the causes and effects of variolae vaccinae, a disease discovered in some western countries of England, particularly Gloucestershire, and known by the name of cow pox. Reprinted by Cassell, 1896, London

26. Kozak W, Zheng H, Conn CA, Soszynski D, van der Ploeg LH, Kluger MJ (1995) Thermal and behavioral effects of lipopolysaccharide and influenza in interleukin-1 beta-deficient mice. Am J Physiol 269: 969–977

27. Lane JM, Ruben FL, Neff JM, Millar JD (1969) Complications of smallpox vaccination, 1968. National surveillance in the United States. N Engl J Med 281: 1 201–1 208

28. Liptakova H, Kontsekova E, Alcamí A, Smith GL, Kontsek P (1997) Analysis of an interaction between the soluble vaccinia virus-coded type I interferon (IFN)-receptor and human IFN-alpha1 and IFN-alpha2. Virology 232: 86–90

29. Mackett M, Smith GL, Moss B (1982) Vaccinia virus: a selectable eukaryotic cloning and expression vector. Proc Natl Acad Sci USA 79: 7 415–7 419

30. Mackett M, Smith GL, Moss B (1984) General method for production and selection of infectious vaccinia virus recombinants expressing foreign genes. J Virol 49: 857–864

31. Massung RF, Liu LI, Qi J, Knight JC, Yuran TE, Kerlavage AR, Parsons JM, Venter JC, Esposito JJ (1994) Analysis of the complete genome of smallpox variola major virus strain Bangladesh-1975. Virology 201: 215–240

32. McMahan CJ, Slack JL, Mosley B, Cosman D, Lupton SD, Brunton LL, Grubin CE, Wignall JM, Jenkins NA, Branan CI, Copeland NG, Huebner K, Croce CM, Cannizzaro LA, Benjamin D, Dower SK, Spriggs MK, Sims JE (1991) A novel IL-1 receptor, cloned from B cells by mammalian expression, is expressed in many cell types. EMBO J 10: 2 821–2 832

33. Morikawa S, Ueda Y (1993) Characterization of vaccinia surface antigen expressed by recombinant baculovirus, Virology 193: 753–761

34. Moss B (1996) Poxviridae: the viruses and their replication. In: Fields BN, Knipe DM, Howley PM (eds) Fields Virology. Lippincott-Raven Press, New York, vol 2, pp 2 637–2 671

35. Moss B, Winters E, Coopper JA (1981) Deletion of a 9,000-base-pair segment of the vaccinia virus genome that encodes nonessential polypeptides. J Virol 40: 387–395

36. Mossman K, Nation P, Macen J, Garbutt M, Lucas A, McFadden G (1996) Myxoma virus M-T7, a secreted homolog of the interferon-gamma receptor, is a critical virulence factor for the development of myxomatosis in European rabbits. Virology 215: 17–30

37. Mossman K, Upton C, Buller RM, McFadden G (1995) Species specificity of ectromelia virus and vaccinia virus interferon-gamma binding proteins. Virology 208: 762–769

38. Novick D, Cohen B, Rubinstein M (1994) The human interferon a/b receptor: characterization and molecular cloning. Cell 77: 391–400

39. Panicali D, Paoletti E (1982) Construction of poxviruses as cloning vectors: insertion of the thymidine kinase gene from herpes simplex virus into the DNA of infectious vaccinia virus. Proc Natl Acad Sci USA 79: 4 927–4 931

40. Patel AH, Gaffney DF, Subak-Sharpe JH, Stow ND (1990) DNA sequence of the gene encoding a major secreted protein of vaccinia virus, strain Lister. J Gen Virol 71: 2 013–2 021

41. Premack BA, Schall TJ (1996) Chemokine receptors: gateways to inflammation and infection. Nature Med 2: 1 174–1 178

42. Shchelkunov SN, Massung RF, Esposito JJ (1995) Comparison of the genome DNA sequences of Bangladesh-1975 and India-1967 variola viruses. Virus Res 36: 107–118

43. Smith CA, Smith TD, Smolak PJ, Friend D, Hagen H, Gernart M, Park L, Pickup DJ, Torrance D, Mohler K, Schooley K, Goodwin RG (1997) Poxvirus genomes encode a secreted, soluble protein that preferentially inhibits β chemokine activity yet lacks sequence homology to known chemokine receptors. Virology 236: 316–327

44. Smith GL, Chan YS (1991) Two vaccinia virus proteins structurally related to interleukin-1 receptor and the immunoglobulin superfamily. J Gen Virol 72: 511–518

45. Smith GL, Mackett M, Moss B (1983) Infectious vaccinia virus recombinants that express hepatitis B virus surface antigen. Nature 302: 490–495

46. Smith GL, Symons JA, Alcamí A (1998) Poxviruses; interfering with interferon. Semin Virol 8: 409–418

47. Smith GL, Symons JA, Khanna A, Vanderplasschen A, Alcamí A (1997) Vaccinia virus immune evasion. Immunol Rev 159: 137–154

48. Spriggs M, Hruby DE, Maliszewski CR, Pickup DJ, Sims JE, Buller RML, Vanslyke J (1992) Vaccinia and cowpox viruses encode a novel secreted interleukin-1 binding protein. Cell 71: 145–152

49. Symons JA, Alcamí A, Smith GL (1995) Vaccinia virus encodes a soluble type I interferon receptor of novel structure and broad species specificity. Cell 81: 551–560

50. Ueda Y, Ito M, Tagaya I (1969) A specific surface antigen induced by poxvirus. Virology 38: 180–182

51. Ueda Y, Morikawa S, Matsuura Y (1990) Identification and nucleotide sequence of the gene encoding a surface antigen induced by vaccinia virus. Virology 177: 588–594

52. Ueda Y, Tagaya I (1973) Induction of skin resistance to vaccinia virus in rabbits by vaccinia-soluble early antigens. J Exp Med 138: 1 033–1 043

53. Ueda Y, Tagaya I, Amano H, Ito M (1972) Studies on the early antigens induced by vaccinia virus. Virology 49: 794–800

54. Upton C, Mossman K, McFadden G (1992) Encoding of a homolog of IFN-γ receptor by myxoma virus. Science 258: 1 369–1 372

55. Uzé G, Lutfalla G, Gresser I (1990) Genetic transfer of α functional human interferon α receptor into mouse cells: cloning and expression of its cDNA. Cell 60: 225–234

Authors' address: Prof. G. L. Smith, Sir William Dunn School of Pathology, University of Oxford, South Parks Road, Oxford OX1 3RE, U.K.

Learning from our foes: a novel vaccine concept for influenza virus

P. Palese[1], **T. Muster**[2], **H. Zheng**[1], **R. O'Neill**[1] and **A. Garcia-Sastre**[1]

[1]Department of Microbiology, Mount Sinai School of Medicine,
New York, New York, U.S.A.
[2]Department of Dermatology, University of Vienna Medical School,
Vienna, Austria

Summary. Concerted efforts to study the molecular biology of influenza viruses and the ability to genetically engineer them have dramatically advanced our understanding of the functions of influenza viral genes and gene products. The only nonstructural protein (NS1) coded for by the influenza virus was shown to possess interferon antagonist activity and thus to play an important role in countering the interferon (antiviral) response of the host following infection. Influenza A and B virus mutants with "weak" anti-interferon activity are highly attenuated because the host is able to mount an effective interferon response. It is suggested that these NS1-modified attenuated influenza viruses can induce a protective immune response and that they are ideal live virus vaccine candidates against influenza.

Introduction

Influenza viruses continue to cause widespread disease in humans and animals. Although the 1918 pandemic appears to have been uniquely devastating, we have experienced three other extraordinary pandemic waves in this century: in 1957, 1968 and 1977. The 1918 pandemic was caused by influenza viruses of the haemagglutinin subtype 1 (H1) and neuraminidase subtype 1 (N1) variety, while the 1957 pandemic was caused by H2N2 viruses, that of 1968 by H3N2 viruses, and that of 1977 by re-emergent H1N1 viruses (Fig. 1). The morbidity and mortality in humans is unfortunately not limited to these pandemic periods, and the toll exacted during interpandemic years can also be quite dramatic [7]. Furthermore, one thing appears to be certain: epidemics and occasional pandemics will continue to occur in the future.

For these reasons, attempts have been made to therapeutically intervene and/or to prophylactically treat influenza. Amantadine and rimantidine have been approved for medical treatment, and neuraminidase inhibitors are currently being developed as effective antivirals against influenza A and B viruses. Although

132   P. Palese et al.

Epidemiology of human influenza
viruses

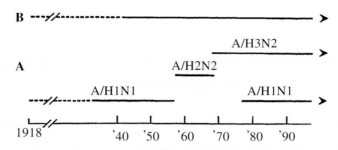

Fig. 1. Influenza A and B viruses circulating in humans during the 20th century. Broken lines indicate that the virus samples are not available for these periods. Influenza A viruses of the H1N1, H2N2 and H3N2 subtypes were identified during time periods as indicated

such antiviral compounds are primarily designed for the treatment of influenza, they may also have the potential for prophylaxis. However, a more general and possibly more cost-effective approach for the prevention of influenza would be the use of safe and effective vaccines. [7, 16].

Current vaccine strategies against influenza

The vaccines currently licensed for humans are of the inactivated (killed) type [16]. They contain three components: two influenza A virus types (H1N1 and H3N2) and one influenza B virus. Most of these vaccine preparations consist of reassortants deriving the genes coding for the haemagglutinin (HA) and neuraminidase (NA) from recently circulating strains and the remaining genes from a high-yield virus such as A/PR/8/34 virus [9]. The vaccine formulations are thus trivalent and administration by the parenteral route induces mostly a systemic antibody response. Carefully controlled studies have shown an efficacy as high as 60–80%, depending on the patient population which was vaccinated and on the vaccine preparations used for the trials [16]. However, there appears to be room for improvement, both in terms of ease of administration (i.e. injection vs. a nasal spray) and of the degree of protection afforded by the vaccine formulation. The most successful approach so far has been the development of live influenza virus vaccines based on cold-adaptation. The cold-adapted (ca) strain A/Ann Arbor/6/60 prepared by John Maassab [13] was developed by passaging the wild type virus in primary chick kidney cells at 25 °C, which is a suboptimal temperature for wild type virus. The resulting ca donor strain virus (i) was temperature sensitive above 38 °C and (ii) was attenuated for humans, mice, and ferrets. Although the molecular basis for attenuation is not clearly understood for the ca virus, the temperature sensitivity of the ca strain restricts the virus to growth in the upper respiratory tract. The immunogenicity of this ca strain, as well as of the reassortants with varying surface protein genes, is adequate to induce protective immunity against experimental or natural challenge with wild type influenza viruses. The suggested use for the ca virus vaccine is for pediatric populations in which the inactivated vaccine has been shown less effective [16].

Additional vaccination approaches concern the induction of neuraminidase-inhibiting antibodies through the administration of viral neuraminidases [8] and immunization using DNA [3]. The latter method of producing viral proteins in the host may provide some of the advantages associated with live virus vaccines. However, it is unlikely to be effective in inducing protective local mucosal immunity, which is one of the hallmarks of a nasally applied live influenza virus vaccine.

We are proposing an alternative approach which derives from our molecular studies of the virus and which is based on the generation of attenuated influenza viruses by reverse genetics methods.

Functions of the influenza virus genes

In recent decades, much progress has been made on characterizing the RNA and protein structure of the virus, and much has been learned about the unique mechanisms of influenza virus gene expression. There are nine structural proteins coded for by the virus, and at least four of them (PB1, PB2, PA and NP) are associated with RNA transcription/replication. The HA, NA and the ion channel protein M2 are inserted into the lipid membrane of the virus and the ribonucleoprotein core (RNP) contains the polymerase complex as well as the matrix (M1) protein and the nuclear export protein (NEP). The only nonstructural protein is the NS1 protein (Table 1).

In the past, several functions and characteristics have been described for the NS1 protein (Table 2). Specifically, it was suggested that the NS1 blocks host cell mRNA polyadenylation, inhibits the nuclear export of host mRNAs, inhibits pre-mRNA splicing and has an effect on translation of viral mRNAs [10, 11]. In addition, several cellular proteins have been identified which interact with the influenza virus NS1 protein [2, 10, 14, 20, 21]. Although some of these properties of the NS1 protein may contribute to the biological characteristics of influenza viruses in infected cells, recent experiments suggest that the NS1 protein of influenza viruses is not required for replication. Rather, the NS1 appears to be an accessory protein which functions as an interferon antagonist [5].

The NS1 gene of influenza A viruses is an accessory gene

Taking advantage of reverse genetics techniques, it was possible to generate an influenza virus which lacked the open reading frame of the NS1 protein [5] (Fig. 2). This delNS1 virus is (i) able to grow in Vero cells which are compromised in their ability to mount an interferon response. Also, (ii) the delNS1 virus is able to grow in STAT1 knock-out mice and to kill these transgenic mice following intranasal inoculation with 5×10^4 pfu within a period of seven days. STAT1-/- mice are unable to express the STAT1 protein which is essential for the signal transduction of the interferon type I and type II pathways. Finally, (iii) in 293 cells, the delNS1

Table 1. Functions of influenza A virus proteins

Encoded protein	Predicted molecular weight[a]	Functions
PB1	86,500	Polymerase subunit, RNA elongation
PB2	85,700	Polymerase subunit, binds host cell mRNA cap, endonuclease
PA	82,400	Polymerase subunit
HA	61,468[b]	Binds cell surface receptor, fusion protein; possesses neutralizing epitopes
NP	56,101	RNA binding protein, functions in replication, RNA trafficking
NA	50,087[c]	Neuraminidase, promotes virus release, prevents viral aggregate formation
M1	27,801	Forms viral matrix, regulates RNP trafficking into and out of nucleus
NS1	26,815	Interferon antagonist, possible role in viral gene expression
M2	11,010	Ion channel
NEP(NS2)	14,216	Nuclear export factor

[a]Predicted molecular weight of unmodified, monomeric proteins of influenza A/PR/8/34 virus

[b]HA is a glycoprotein which forms trimers. A precursor form, HA0, is cleaved into two disulfide-linked subunits, HA1 and HA2

[c]NA is a glycoprotein which forms tetramers. Modified from [1]

Table 2. Proposed functions of the influenza A virus NS1 protein

- Inhibition of host mRNA polyadenylation
- Inhibition of pre-mRNA splicing
- Inhibition of nuclear export of mRNA
- Regulation of viral RNA polymerase activity
- Enhancement of translation of viral mRNAs
- Inhibition of PKR binding of RNA
- Interferon antagonist

virus induces higher levels of interferon than wild type virus, as measured by the stimulation of transcription of a reporter gene from an interferon-sensitive promoter. We conclude from these data [5] that the NS1 protein is not essential for the virus to replicate in interferon-deficient cells, and that it is therefore a quintessential accessory protein. However, the NS1 is required as an interferon antagonist if the virus is to "win" against a host cell with an intact interferon system (Fig. 3).

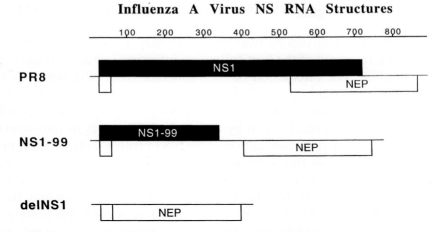

Fig. 2. The influenza A virus NS segment codes for two proteins, the NS1 and the NEP (NS2). The NS1 gene (filled rectangle) of influenza A/PR/8/34 and of the isogenic NS1-99 virus codes for a protein of 230 and 99 amino acids, respectively . The open reading frame of the NEP (open rectangle) is the same for influenza A/PR/8/34, NS1-99 and delNS1 viruses. In the case of PR8 and NS1-99 viruses, the NEP is coded for by spliced mRNAs. The delNS1 virus lacks the open reading frame of an NS1 protein

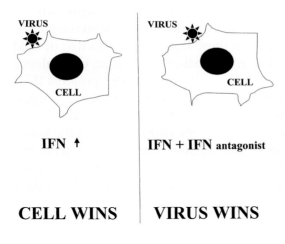

Fig. 3. Influenza virus lacking an NS1 gene or coding for a "weak" NS1 protein (interferon antagonist) is not able to counter the interferon response of the host cell. The antiviral (interferon) response of the host results in protection of the host (left panel). Influenza virus with a "strong" NS1 protein can overcome the interferon response of the host (right panel)

Attenuation of influenza viruses via modified NS1 genes

We studied the replication of influenza virus A/WSN/33 in STAT1 knock-out mice following intranasal inoculation, and found that this virus shows about a 100-fold lower LD_{50} in these mice than in wild type mice. Furthermore, the transgenic knock-out mice developed a systemic infection, with virus found to replicate in the spleen, liver, brain and kidney; in wild type (interferon-competent) C57B6 mice replication was restricted to the lung under the conditions used [4]. We interpret these data to mean that the presence of an interferon type I signalling pathway severely limits replication of influenza viruses, even in the presence of a full length NS1 protein. By the same token, a virus with a highly efficient NS1 protein may overcome this restriction imposed by the host and therefore be more virulent.

The potential mechanisms by which the NS1 may achieve the neutralization of the interferon response of the host following infection includes (i) the binding of NS1 to dsRNA. The interaction of the influenza virus NS1 with RNA was first recognized by Yoshida et al. [22] and this characteristic of the protein, i.e. cleansing of the cell of interferon-inducing dsRNA (or highly structured ssRNA) may be the predominant mechanism by which the NS1 exerts its anti-interferon activity. Alternatively, (ii) the interaction of the NS1 protein with the PKR protein kinase may interfere with the interferon-induced antiviral defense system. By expressing the NS1 protein (an early protein), influenza viruses may modulate the interferon response of the host [6, 12, 18]. A third way (iii) in which the NS1 affects the interferon response may involve activation of host cell proteins. For example, influenza virus infection has been shown to activate the host cell protein p58(IPK), which is an inhibitor of the double-stranded RNA-activated protein kinase (PKR) [15].

The absence of the NS1 protein in a delNS1 virus-infected cell thus results in the unmitigated ability of the cell to mount an antiviral response (Fig. 3). The consequence is that the virus will not be able to replicate efficiently and cause disease. In fact, the highest dose of delNS1 virus (10^6 PFU) did not result in the death of wild type mice. A virus expressing an NS1 protein with only 99 amino acids (rather than the 230 aa of the full length protein) showed intermediate virulence in mice, with an LD_{50} of at least three logs higher than that of wild type virus (unpublished results). Thus, we can now study a spectrum of virulence characteristics of isogenic viruses, which differ only in their NS1 proteins (Fig. 2). In the case of the intermediate length (truncated) NS1 protein, we suggest that the stability of the NS1 is compromised (or its affinity for its substrate is altered), resulting in the reduced ability of the NS1-99 virus to counter the host's interferon response. Similar differences in attenuation characteristics were observed between influenza B/Yam/1/73 virus and its isogenic B virus derivatives B/201[17] and AWBY-234 [19]. These viruses differ only in the lengths of their NS1 proteins.

Possible use as vaccine strains of influenza viruses with defects in the NS1 protein

Most of the successful human live virus vaccines are host range mutants, which were selected by repeated passaging in tissue culture or animal hosts. The resulting strains had a lower virulence in humans than the wild type viral strains. This Jennerian approach was successful for many viral vaccines, including those against measles, mumps, rubella and chicken pox. However, in most instances, the molecular basis for attenuation of the vaccine strains is not clearly understood.

We suggest that mutating the NS1 gene of influenza viruses leads to a reduction in the anti-interferon activity, and thus represents a rational approach to obtain live attenuated vaccine strains. We postulate that such strains will replicate to high enough titers in the human host to induce a protective immune response, including a mucosal response without causing disease. Preliminary data suggest that immunization of mice with the delNS1 virus as well as with the NS1-99 virus resulted in complete protection against challenge four weeks later with $100\,LD_{50}$ of wild type virus (unpubl.). Although human vaccine trials are the only way to judge the safety and efficacy of such a protocol, we are encouraged by the fact that type I interferons are redundant and that it is therefore likely that all humans will be able to mount an adequate interferon response to influenza viruses with mutated NS1 genes. We also suggest that changing the HA and NA genes of NS1-modified strains (to reflect the current and future circulating strains) will not alter the "weak" interferon antagonist activity of the vaccine strains. This assumption is predicated on the fact that the glycoproteins of influenza viruses (HA and NA) do not have interferon antagonist activity and thus should not affect the attenuation characteristics of the NS1-modified virus.

In conclusion, we believe that the NS1 protein of influenza A and B viruses possesses anti-interferon activity. Influenza virus strains with a "weak" NS1 protein would be subject to a vigorous interferon response by the human host and thus are unlikely to cause disease. Such attenuated strains are predicted to be safe and effective vaccine strains against influenza.

References

1. Basler C, Palese P (1999) Influenza viruses. In: Lederberg J (ed) Encyclopedia of microbiology. Academic Press, San Diego (in press)
2. Chen Z, Li Y, Krug RM (1999) Influenza A virus NS1 protein targets poly (A)-binding protein II of the cellular 3′-end processing machinery. EMBO J 18 : 2273–2283
3. Donnelly JJ, Ulmer JB, Liu MA (1998) DNA vaccines. Dev Biol Standards 95: 43–53
4. Garcia-Sastre A, Durbin RK, Zheng H, Palese P, Gertner R, Levy DE, Durbin JE (1998b) The role of interferon in influenza virus tissue tropism. J Virol 72: 8 550–8 558
5. Garcia-Sastre A, Egorov A, Matassov D, Brandt S, Levy DE, Durbin JE, Palese P, Muster T (1998a) Influenza A virus lacking the NS1 gene replicates in interferon-deficient systems. Virology 252: 324–330
6. Hatada E, Saito S, Fukuda R (1999) Mutant influenza viruses with a defective NS1 protein cannot block the activation of PKR in infected cells. J Virol 73: 2 425–2 423
7. Hayden FG, Palese P (1997) Influenza virus. In: Richman DD, Whitley RJ, Hayden FG (eds) Clinical virology. Churchill Livingstone, New York, pp 911–942
8. Johansson BE, Matthews JT, Kilbourne ED (1998) Supplementation of conventional influenza A vaccine with purified viral neuraminidase results in a balanced and broadened immune response. Vaccine 16: 1009–15
9. Kilbourne ED, Schulman JL, Schild GC, Schloer G, Swanson J, Bucher D (1971) Related studies of a recombinant influenza virus vaccine. I. Derivation and characterization of virus and vaccine. J Infect Dis 124: 449–462
10. Krug RM (1999) Unique functions of the NS1 protein. In: Nicholson KG, Webster RG, Hay AJ (eds) Textbook of influenza. Blackwell, Oxford, pp 82–92
11. Lamb RA, Krug RM (1996) Orthomyxoviridae: The viruses and their replication. In: Fields BN, Knipe DM, Howley PM (eds) Virology, 3rd ed. Lippincott-Raven, Philadelphia, pp 1 353–1 395
12. Lu Y, Wambach M, Katze MG, Krug RM (1995) Binding of the influenza virus

NS1 protein to double-stranded RNA inhibits the activation of the protein kinase that phosphorylates the elF-2 translation initiation factor. Virology 214: 222–228

13. Maassab HF, LaMontagne JR, DeBorde DC (1998) Live influenza virus vaccines. In: Plotkin SA, Mortimer EA (eds) Vaccines. Saunders, Philadelphia, pp 435–457
14. Marion RM, Fortes P, Beloso A, Dotti C, Ortin J (1999) A human sequence homologue of Staufen is an RNA-binding protein that is associated with polysomes and localizes to the rough endoplasmic reticulum. Mol Cell Biol 19: 2 212–2 219
15. Melville MW, Tan SL, Wambach M, Song J, Morimoto RI, Katze MG (1999) The cellular inhibitor of the PKR protein kinase, p58 (IPK), is an influenza virus-activated co-chaperone that modulates heat shock protein 70 activity. J Biol Chem 274: 3 797–803
16. Murphy BR, Webster RG (1996) Orthomyxoviruses In: Fields BN, Knipe DM, Howley PM (eds) Virology, 3rd ed. Lippincott-Raven, Philadelphia, pp 1 397–1 445
17. Norton GP, Tanaka T, Tobita K, Nakada S, Buonagurio DA, Greenspan D, Krystal M, Palese P (1987) Infectious influenza A and B virus variants with long carboxyl terminal deletions in the NS1 polypeptides. Virology 156: 204–13
18. Tan SL, Katze MG (1998) Biochemical and genetic evidence for complex formation between the influenza A virus NS1 protein and the interferon-induced PKR protein kinase. J Interferon Cytokine Res 18: 757–66
19. Tobita K, Tanaka T, Odagiri T, Tashiro M, Feng S-Y (1990) Nucleotide sequence and some biological properties of the NS gene of a newly isolated influenza B virus mutant which has a long carboxyl terminal deletion in the NS1 protein. Virology 174: 314–319
20. Wolff T, O'Neill RE, Palese P (1996) Interaction cloning of NS1-I, a human protein that binds to the nonstructural NS1 proteins of influenza A and B viruses. J Virol 70: 5 363–72
21. Wolff T, O'Neill RE, Palese P (1998) NS1-binding protein (NS1-BP): A novel human protein that interacts with the influenza A virus nonstructural NS1 protein is relocalized in the nucleus of infected cells. J Virol 72: 7 170–7 180
22. Yoshida T, Shaw MW, Young JF, Compans RW (1981) Characterization of the RNA associated with influenza A cytoplasmic inclusions and the interaction of NS1 protein with RNA. Virology 110: 87–97

Authors' address: Dr. P. Palese, Department of Microbiology, Box 1124, Mount Sinai School of Medicine, 1 Gustave L. Levy Place, New York, NY 10029, U.S.A.

Pathogenic aspects of measles virus infections

S. Schneider-Schaulies and **V. ter Meulen**

Institute of Virology and Immunobiology, University of Würzburg, Würzburg, Germany

Summary. Measles virus (MV) infections normally cause an acute self limiting disease which is resumed by a virus-specific immune response and leads to the establishment of a lifelong immunity. Complications associated with acute measles can, on rare occasions, involve the central nervous system (CNS). These are postinfectious measles encephalitis which develops soon after infection, and, months to years after the acute disease, measles inclusion body encephalitis (MIBE) and subacute sclerosing panencephalitis (SSPE) which are based on a persistent MV infection of brain cells. Before the advent of HIV, SSPE was the best studied slow viral infection of the CNS, and particular restrictions of MV gene expression as well as MV interactions with neural cells have revealed important insights into the pathogenesis of persistent viral CNS infections. MV CNS complication do, however, not large contribute to the high rate of mortality seen in association with acute measles worldwide. The latter is due to a virus-induced suppression of immune functions which favors the establishment of opportunistic infections. Mechanisms underlying MV-mediated immunosuppression are not well understood. Recent studies have indicated that MV-induced disruption of immune functions may be multifactorial including the interference with cytokine synthesis, the induction of soluble inhibitory factors or apoptosis and negative signalling to T cells by the viral glycoproteins expressed on the surface of infected cells, particularly dendritic cells.

Introduction

Measles still ranks as one of the leading causes of childhood mortality world-wide. In developing countries there are 30–40 million cases of measles reported each year and 1–2 million deaths are associated with measles virus (MV) infection. In 1995 measles was listed among the ten most common and ten most deadly diseases world-wide. Although measles was largely controlled due to the availability of an efficient live vaccine in industrialised nations, there was no such success in Third World countries. Moreover, measles outbreaks have been documented in industrialised countries with high vaccine coverage, such as the United States, Canada and Germany, where many cases had a history of measles vaccination as documented by seroconversion. As there is no evidence for the spread of antigenic variants of measles virus that could escape vaccine-induced immunity, vaccination, unlike naturally acquired measles, may not induce lifelong immunity to reinfection.

The diseases

Measles virus (MV) is spread by aerosol and initially replicates in the epithelial cells of the respiratory tract. After spread to and amplification in local lymph nodes, a viraemia leads to the dissemination of the virus to multiple lymphoid tissues and other organs, including skin, liver and the gastrointestinal tract. The viraemia is cell-associated and MV can be detected in peripheral blood mononuclear cells (PBMCs) [20, 40, 59]. An efficient immune activation towards MV is observed in the course of acute measles that leads to viral clearance and the establishment of life-long immunity against reinfection (reviewed in [24]). Activation of T-cell immunity appears to be essential for overcoming acute measles, whereas the efficient induction of a humoral virus-specific immune response is a prerequisite for protection against reinfection. Infection of PBMCs, as well as a pronounced lymphopaenia, has been suggested as major causes of the suppression of immune functions observed during acute measles and lasting for a number of weeks after recovery [10]. On the basis of this immunosuppression, complications such as diarrhoea, pneumonia, laryngotracheobronchitis, otitis media and stomatitis may arise; these have a particularly high case fatality rate in developing countries [31]. Acute postinfectious measles encephalitis develops during or shortly after acute measles (approximately 0.1% of cases, with a mortality rate of 10–30%), likely as a consequence of a measles virus induced autoimmune reaction against brain antigens.

Late complications associated with acute measles are rare. These include two fatal diseases of the CNS: subacute sclerosing panencephalitis (SSPE) and, in immunocompromised individuals, measles inclusion body encephalitis (MIBE) [39, 45]. Both diseases develop because of persistent MV infections in brain cells. Giant cell formation and budding virus particles as typically found in measles infection, are virtually absent in SSPE and MIBE brains, indicating a defective MV replication in CNS tissue. Pathognomonic for SSPE, which develops years to months after acute measles, are exceptionally high antibody titres to the majority of MV structural proteins in serum and CSF specimens (excluding the matrix (M) protein). The presence of this MV-specific oligoclonal IgG in the CSF of the patients reflects a state of CNS hyperimmunisation and results from a local production of antiviral antibodies by sensitised lymphocytes that have invaded this compartment. In measles inclusion body encephalitis, no such hyperimmune response to MV is found as other systemic diseases generally impair the immune system.

MV organisation and protein functions

MV particles consist of a lipid envelope surrounding the viral RNP complex, which is composed of genomic RNA associated with nucleocapsid (N) protein and the viral polymerase complex (L protein and the phosphorylated cofactor, P protein) (Fig. 1A). Both viral transmembrane proteins (fusion (F) and haemagglutinin (H) proteins) are present on the envelope surface. It is the aminoterminus of the H protein that protrudes through the cytoplasmic and viral membranes (type II glycoprotein), while the F protein is anchored near the carboxyterminus (type I glycoprotein). One or both of the cytoplasmic domains are believed to interact with the matrix (M) protein which, in turn, forms the link to the RNP core structure.

The viral genome is a nonsegmented RNA molecule of negative polarity that is about 16 kb in length [27] (Fig. 1B). The genome encodes six structural genes for which the reading frames are arranged linearly and without overlap in the following order: 3′ nucleocapsid protein (N, 60 kD), phosphoprotein (P, 70 kD), matrix protein (M, 37 kD), fusion protein (F, disulfide linked 41 kD F_1 and 20 kD F_2 proteins, cleavage products of a 60 kD precursor F_0 protein), haemagglutinin protein (H, 80 kD, existing as disulfide linked homodimer) and the large protein (L, 220 kD) on its 5′end. The genome is flanked by noncoding 3′leader and 5′trailer sequences that contain specific encapsidation signals and the promoters used for replication of the viral genome by the polymerase complex (P and L proteins). Within the 3′leader sequence, transcription of the viral monocistronic mRNAs from the encapsidated genome is initiated. The coding regions of the viral genome are interspersed by conserved intergenic regions containing a polyadenylation signal, a central conserved trinucleotide and a reinitiation signal. From the P gene, three nonstructural proteins, C, R and V are expressed. Since V is expressed from edited P transcripts (with an editing frequency of about 50%), it shares a common amino-terminal domain with P, while a zinc-finger like domain is present on its carboxy-terminus.

Both viral glycoproteins, the fusion protein F and the haemagglutinin protein H, are required for efficient fusion after viral attachment [44, 75]. Both proteins are important antigens for the induction of virus-neutralising antibodies. The H protein can be isolated as a tetrameric complex from the cell membrane and its ability to agglutinate red blood cells from sheep and monkeys, but not humans, has long been recognised. Glycosylation has been shown essential for haemadsorption, probably by stabilising the highly complex tertiary structure of the protein. The strict structural constraints of this protein have so far prevented the identification of domains essential for the interaction with the receptor complex, which consists of CD46 [16, 41] physically associated with moesin [17, 18, 55]. Both CD46 and moesin reveal a wide tissue distribution in vivo [29] although a restricted expression of only certain isoforms of CD46 has been detected on brain cells [11]. H protein mediates attachment to CD46 and, at least for H proteins predominantly of vaccine strains, subsequently downregulates CD46 from the cell surface [32, 56–58]. Functional consequences of removal of CD46 proteins from the cell surface include an increased vulnerability of the infected cells to complement lysis [58, 67]. Amino acids 451 and 481, and to some degree, 211 and 243, within the H sequence have been shown to be essential for this downregulation [7, 35]. As indicated by recent studies, lymphotropic MV wildtype strains may exhibit a

A.

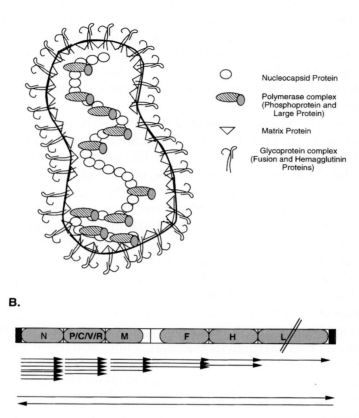

B.

Fig. 1. Schematic representation of the MV particle, MV genome organisation and RNA synthesis. **A** The pleomorphic virus consists of two major subunits. The viral genome is tightly encapsidated by the nucleocapsid protein, and the polymerase complex consisting of the polymerase or large protein (*L*) together with the cofactor protein, the phosphoprotein (*P*), are also associated. The viral core is surrounded by a host cell derived lipid envelope in which the two viral glycoproteins, the fusion (*F*) and the haemagglutinin (*H*) protein are inserted. The matrix protein (*M*) is thought to interact with both the viral core and the envelope. **B** The genome of MV is a single stranded nonsegmented RNA molecule of negative polarity. The viral genes are sequentially arranged, flanked by noncoding leader or trailer sequences and interspersed by conserved intergenic sequences. Except for the P gene, all MV genes are monocistronic. Within the P gene, the viral nonstructural proteins V, C and R are also encoded. The viral polymerase complex sequentially transcribes mono- and polycistronic (not shown) polyadenylated mRNAs along the viral genome which accumulate with decreasing efficiency. Following primary transcription and translation, a full length replicative intermediate of the genomic RNA is produced which serves as a template for the replication of the minus strand genome

differential receptor usage [8, 28], the MV wildtype receptor is, however, not yet known. As revealed by transfection experiments, H also exerts a helper function in F-mediated membrane fusion [14, 75]. However, under certain circumstances, such as at high expression levels, fusion can be mediated by F alone in the absence of H. The precursor protein F_0 is cleaved in the Golgi compartment by a subtilisin-like protease into two disulfide-linked subunits, F_1 and F_2 [74]. The fully processed $F_{1/2}$ protein is incorporated into the cell membrane as an oligomer.

MV gene expression in persistent brain infections

As infectious virus is usually absent from brain material of patients with subacute sclerosing panencephalitis (SSPE) and measles inclusion body encephalitis, MV gene expression is restricted in a way that allows intracellular survival and replication of the virus while remaining inaccessible to the host immune system. This is generally achieved by abolishing viral gene functions associated with maturation and budding, and maintaining those required for transcription and replication [5, 9, 15, 50, 60] (Fig. 2). Whereas N and P proteins can usually be detected in infected brain cells, M, F and H proteins are, if at all, barely expressed only in a small percentage of these cells [36]. The molecular basis of the restrictions was analysed using either viral RNAs directly isolated from brain material of patients with per-

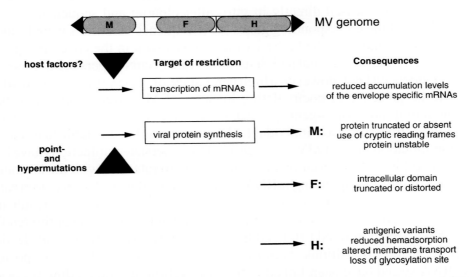

Fig. 2. Summary of the restrictions of MV gene expression in SSPE. In persistent MV CNS infection, the expression of the envelope genes M, F and H is restricted both on transcriptional and translational level. Most likely due to the activity of as yet undefined host factors, MV-envelope protein specific transcripts accumulate to very low levels in infected brain cells, and, occasionally, high levels of bicistronic transcripts are produced. As revealed by sequence analysis, mutations (point mutations due to the infidelity of the viral polymerase and hypermutations due to the activity to a cellular enzyme complex) accumulate in the corresponding reading frames which completely abolish the expression of the envelope proteins or direct the synthesis of envelope proteins with altered functions

sistent CNS infections, or alternatively, using tissue cells persistently infected with MV, either with standard MV strains or with so called 'SSPE isolates' obtained by cocultivation of SSPE brain material [3, 4, 9, 12, 13, 33, 50, 53, 54, 60, 76].

Transcriptional regulation

Because of the viral transcription strategy, the accumulation levels of the MV-specific mRNAs drop with their distance from the 3′end of the genome. This creates a shallow gradient for the expression of MV genes distal to the N gene in a lytic infection in nonneural tissue culture cells [12]. Quantitative analyses of the steady state levels in brain tissue typically revealed a much steeper decrease in the frequency of MV specific monocistronic mRNAs [12,13] resulting in an underrepresentation of those encoding the viral envelope proteins M, F and H. Qualitiative alterations in the MV transcription pattern such as an enhanced frequency of polycistronic transcripts have also been occasionally found, which led, in some instances, to the complete replacement of adjacent monocistronic mRNAs [5,12].

Alterations of viral protein expression and functions

The properties of MV-specific proteins expressed from mRNAs present in infected brain tissue at low levels or more abundantly in persistently infected tissue culture cells, have been assessed directly in vitro after cloning and sequencing of full length copies of the corresponding genes [53]. Sequence variations were found to be widely distributed but the highest variation was seen in the envelope specific genes, initially suggesting that a high mutation rate occurs in persistent infections [15]. As evidenced later, however, a high percentage of these variations more likely reflected the presence of cocirculating lineages of MV strains that might have been the initial infectious agents [3, 50]. Based on a large number of sequences, evolutionary trees for each individual MV gene have been established and the variation/mutation rates of MV genes derived from persistent infections have been re-evaluated [50]. Most variations were not likely to interfere with the biological properties of the corresponding mRNAs as they were silent or led to conservative amino acid exchanges. However, single point mutations that could result in a functional impairment by preterminating or completely abolishing entire reading frames have been described [9]. In addition, biased hypermutation events leading mostly to clustered transitions of uridine (U) to cytosine (C) residues in the positive sense RNA strand have been linked to the inactivation of viral reading frames in SSPE isolates (see below).

Consistent with their key functions in maintaining transcription and replication of the viral genomic RNA during persistent infections, the reading frames of the MV N and P proteins do not harbour gross mutational alterations [3, 73]. Information on variations within the viral L gene is scarce. The sequences of the reading frames of each two SSPE- and wild-type derived L genes were found to be well conserved with the degree of variability being the lowest of all MV structural proteins [33].

In agreement with earlier findings that, in many SSPE sera and CSF samples, antibodies to the M protein were underrepresented, the expression of M protein was affected in the majority of cases. M gene expression alterations in persistent human brain infections are brought about by a number of different mechanisms, including the low levels of mRNA seen in all of the SSPE cases. Disruptions of the reading frame based on single point mutations such as premature termination of translation, the usage of cryptic reading frames and the synthesis of highly unstable translation products have been observed (reviewed in [9, 60]). Within the M gene coding sequences clustered transitions due to hypermutation events were also found which completely inactivated the reading frames [4, 15]. As confirmed in the latter case, the M gene sequences isolated differed in their degree of hypermutation. It appeared that these sequences resulted from sequential hypermutation events of a wild-type-like M gene that could, although with low frequency, still be isolated from SSPE brain material and could be expressed in vitro [4]. These findings suggested that the M gene function of the initial infecting MV strain was nondefective, and its inactivation was apparently not required for the establishment of a persistent brain infection.

Since MV is strictly cell associated during persistent brain infection, no apparent constraints would be expected to maintain the reading frames for the MV glycoproteins F and H. Surprisingly, the sequences of both genes revealed rather low mutation rates [50]. In view of the high degree of overall conservation, it appears even more remarkable that the short cytoplasmic domain of the F proteins isolated from SSPE brain tissue and two SSPE derived cell lines were altered by point mutations or deletions leading either to premature termination or frameshifting within the corresponding reading frame [54]. The putative membrane anchor domain and the cleavage site between the F_1 and F_2 subunits were highly conserved. For the H proteins, alterations of the glycosylation sites have been defined in addition to other mutations including a hypermutation event within the H protein isolated from an SSPE-derived isolate [14].

Functional consequences of mutational alterations of MV F and H variants have been tested by expression of the isolated gene [14]. For all but one F variant, expression, processing and fusogenic activity, when coexpressed with the Edmonston strain (ED) H protein apparently were normal. For the mutated H proteins, addition of complex oligosaccharides and dimerization required for an efficient transport to the cell surface were more or less altered. Moreover, some of these H proteins were not able to cooperate in cell fusion when coexpressed with the Edmonston strain (ED) F protein. Because coexpression of two of these mutated H proteins with their homologous F proteins led to extensive cell fusion, specific functional F-H protein interactions are apparently conserved.

Establishment of persistent MV infections

It is evident that wild-type MVs circulating in the corresponding geographical areas are the causative agents infecting the CNS [50] and there is no evidence for a selection of 'neurotropic' MV strains in natural infections. Thus, persistent MV infections are apparently established after natural infection in primarily nonde-fective MV strains. Neither the route of CNS infection nor factors contributing to the establishment and initial maintenance of persistent MV infection in the CNS are fully understood. Studies on the establishment of MV persistence on the whole animal level have been hampered by the facat that only humans and other primates can be infected. Intracerebral infection of adapted MV strains is usually used to study MV pathogenesis in rodents, but it is only in some cases that persistent infections are established. The availability of CD46 transgenic mouse and rat lines elicited hopes for establishing an animal model that would allow a peripheral infection with MV and studies on persistent CNS infections. It became clear, however, that the expression of CD46 could not increase the susceptibility of infection with MV as compared to nontransgenic littermates [42]. Although also of limited suitability, tissue culture systems have been widely used in the past to study the establishment of MV persistence.

Regulation of MV transcription

The underrepresentation of 5′ MV mRNAs specifying the envelope genes is one of the few findings common to all SSPE cases (see above). Analyses in vitro revealed that overall MV specific transcription is substantially reduced in brain cells as indicated by the steady-state levels of N-specific transcripts per infected cell. After primary infection, accumulation of these transcripts was reduced by about 90% as compared to nonneural cells [61, 62]. Differentiation of the neural host cells increased the overall reduction in MV transcription as documented in brain material from experimentally-infected animals and in tissue culture with infected human neuroblastoma cells treated with differentiating compounds [37, 63, 79]. The progressive decrease in viral mRNA accumulation along the gene order was also apparent in tissue culture with neural cells and in brain tissue of experimentally infected animals [37, 62, 63].

Transcriptional attenuation of MV in brain cells may be enhanced by exoge-nous factors such as virus-neutralising antibodies. Transfer of neutralising anti-H-antibodies was followed by a significant down regulation of MV transcription in experimentally infected newborn rats as compared to controls [37]. In tissue culture, a strongly reduced expression of all MV structural proteins was observed a few days after the addition of neutralising antibodies in persistently MV-infected rodent neural cells, but not in Vero (monkey) cells or human lung fibroblasts [6, 48, 64]. Total MV transcriptional efficiency was reduced up to tenfold as early as 24 h after the application of antibodies, whereas the relative frequencies of the 5′mRNAs were unaffected [64]. The signalling pathways involved have not yet been elucidated.

Interferon-dependent attenuation of MV gene functions may contribute to the establishment of persistent infections. This has been directly demonstrated for the interferon-inducible cytoplasmic human MxA protein, which is readily syn-thesised after MV infection in a variety of cultured cells and expressed to high levels in SSPE brain tissue and in monocytes [34, 65, 66]. Constitutive expression of this protein has been linked to transcriptional attenuation of MV in brain cells [65]. As typically observed in SSPE, downregualation of MV transcription in MxA-transfected brain cells affected both the overall efficiency and the relative frequencies of the 5'mRNAs. Remarkably, the same protein expressed in the human monocytic cell line U-937 specifically inhibited the synthesis of the MV glycoproteins in the absence of any detectable transcriptional control [66]. The molecular mechanism of how this protein acts as a host cell specific factor attenuating gene expression of different viruses at various levels has not yet been resolved. Whether the stimulation of an unidentified host cell kinase in MxA transfected brain cells relates to its antiviral activity has not been directly demonstrated [71].

Biological activity of virus-specific transcripts

Many restrictions at the translational level characterised with SSPE brain-derived MV mRNAs were based on sequence mutations leading to premature termination or complete abolition of the corresponding reading frames (see above). Alternative mechanisms controlling viral gene expression may also be operative since in spite of the presence of wild-type M sequences restricted M protein expression was observed in brain tissue of a patient with SSPE [4]. Moreover, M protein could not be translated in vivo or in vitro from mRAN isolated from in the brain of experimentally infected Lewis rats with subacute measles encephalitis (SAME) independent of detectable sequence alterations [63].

The surprising finding of a hypermutated M gene in a case of MIBE in which 50% of the uridine (U) residues was replaced by cytidine (C) [15] was not attributed to the action of the viral polymerase but rather to a cellular enzyme referred to as duplex RNA dependent adenosine deaminase (DRADA). Further support was lent to this assumption by the high degree of homology between all M gene specific sequences isolated from this particular brain area, indicating a single step event rather than independently occurring point mutations introduced by the polymerase. DRADA has been detected in vitro in a variety of cell types including brain cells [19, 49] and converts adenosine (A) to inosine (I) residues within dsRNA templates with a frequency up to 50%, thereby both destabilsing the dsRNA and altering the informational content of the RNA. Two forms of this enzyme have been characterised, one of them being constitutively expressed, the other being inducible by both type I and type II interferon [46]. The latter, in particular, has been implicated in the site-selective editing of mammalian mRNAs of neural origin and the generation of biased hypermutations in viral mRNAs.

Although mostly defined in M genes, hypermutated sequences have also been found in other MV genes [33, 73]. Clear evidence for the presence of DRADA in extracts prepared from MV-infected tissue culture cells has been obtained [26], and both nuclear and cytoplasmic extracts of neural cell lines were found to be active in modifying a synthetic MV dsRNA in vitro [19, 49].

The potential role of DRADA in contributing to the establishment of MV persistence in vivo was elucidated by sequence analyses performed on M genes isolated from different brain regions of a patient with SSPE [4]. A wild-type-like precursor sequence – closely related to the probable initially infecting virus – underwent at least five independent hypermutation events, giving rise to variant sequences all revealing clustered transitions. Within the M sequences evolving from the precursor, one sequence present in all brains region suggesting that a selective advantage led to a clonal expansion of viruses carrying this sequence. Whatever the consequences of the hypermutated MV sequences on the propagation of the virus in brain tissue, the initial attenuating determinant appears to be dependent on the host cell (DRADA) and not on the virus.

MV induced immunosuppression

Paradoxically, immune activation during measles is accompanied by numerous abnormalities of immune functions. Delayed-type hypersensitivity (DTH) skin test responses to recall antigens disappear, and there is an increase in susceptibility to secondary infections (reviewed in [10]). A marked lymphopaenia is also observed affecting both B- and T-cells (of both the CD4 and CD8 type) as are MV-specific RNA and proteins in a limited number of peripheral blood mononuclear cells (PBMCs) [1, 20, 59]. Typically, PBMCs isolated during and up to several weeks after acute measles are largely refractory to proliferation in response to mitogenic, allogeneic and recall antigen stimulation.

Factors and mechanisms underlying MV-induced immunosuppression are still not understood. As a lymphotropic virus, MV replicates in PBMCs and a certain percentage of these cells may be destroyed directly. It has been shown that MV-infection in tissue culture blocks mitogen-dependent proliferation of PBMCs as well as spontaneous proliferation of cell lines of lymphocytic origin [38, 78]. Moreover, MV does not interfere with effector functions acquired before infection [22]. The low frequency of infected cells throughout the acute infection suggested that both the depletion of lymphocytes and the general suppression of immune functions probably result from indirect mechanisms. Soluble inhibitory factors released from MV-infected lymphocytic/monocytic cells have not been unequivocally identified. The role of apoptosis for MV-induced immunosuppression is unclear. Apoptosis after infection of tissue culture cells is not likely to play a major role in immunosuppression in vivo because it was only observed in infected cells [21]. In mice with severe combined immunodefiency (SCID) grafted with human thymic material, apoptosis was observed in uninfected thymocytes, probably as a result of their interaction with MV-infected epithelial cells [2]. Interactions via cell surface molecules between infected and uninfected cells appear particularly attractive for the induction of immunosuppression; this would explain how MV infection of a relatively small number of cells can have a far-reaching effect on the immune response mounted by a large number of uninfected cells. In vitro, the proliferative response of uninfected T cells to mitogen or to a specific antigen was markedly impaired after cocultivation with autologous, MV-infected and UV-irradiated PBLs. The effect was not mediated by the infection of T cells and was completely sensitive to anti-MV serum, indicating that interactions between viral structural proteins and receptors on the surface of the uninfected cells were essentially involved [51]. It is unlikely that interaction between MV H protein and CD46, the cognate MV receptor, plays a major role in this process, because certain MV strains, mostly of wildtype origin, do not efficiently bind to this molecule [8, 28]. These strains, however, have certainly been active in inducing immunosuppression in vivo and very efficiently do so in vitro (see below). For the same reason, down regulation of lipopolysaccharide-induced synthesis of IL-12, a cytokine essential for triggering of TH1 responses, after crosslinking of CD46 on monocytic cells by MV, its natural ligands (C3b and C4b) or CD46-specific antibodies, can only account for immunosuppression under certain circumstances [30].

Surface interaction of MV glycoproteins is necessary and sufficient to induce proliferative inhibition in lymphocytic/monocytic cells

Aiming to investigate how a minority of infected cells could exert a far reaching negative effect on the proliferation of a far greater number of uninfected PBMCs (as found ex vivo for PBMCs isolated from measles patients), an in vitro system was used in which uninfected, mitogen-stimulated peripheral blood lymphocytes (PBLs (in most cases depleted for monocytes/macrophages)(termed 'responder cells', RCs) were cocultured with PBLs (presenter cells (PCs) infected with MV for 48 h in the presence of mitogen (or, for control, mock infected) and subsequently UV-irradiated) [52] (Fig. 3). By this treatment, proliferation of the PCs as well as release of infectious virus from these cells was almost completely abolished.

Using this assay, we observed that mitogen-dependent proliferation of the RCs was impaired in the presence of infected, but not of uninfected PC. The extent of inhibition correlated with the PC/RC ratio and a reduction of 50% was still seen using a PC/RC ratio of 1/100 [52]. In addition to mitogen-driven proliferation, allogeneic as well as CD3-induced stimulation were also affected. The inhibition was observed both with autologous or allogeneic PC/RC cocultures, and major donor-dependent variations were not detectable. The inhibitory effect of MV-infected PCs was not confined to primary lymphocytes but also applied to spontaneous proliferation of human tissue culture cells of both lymphocytic and monocytic origin but not of adherent cells. Since MV-infected PCs also conferred unresponsiveness to mitogen-induced proliferation of mouse and rat lymphocytes, human CD46 (which is not expressed on rodent cells), RC infection (rodent cells are usually resistant to MV infection due to an intracellular block [42]) and PC/RC fusion (which does not occur with rodent cells in the absence of CD46), do not play a major role in this inhibition.

Inclusion of a filter with a pore size of 200 nm (which permits diffusion of mediators) between PCs and RCs completely abolished PC-dependent inhibition suggesting that the effect was dependent on a PC/RC surface contact, and independent of soluble factors. Similarly, PC surface molecules other than viral apparently were essentially not involved because inhibition of RC proliferation

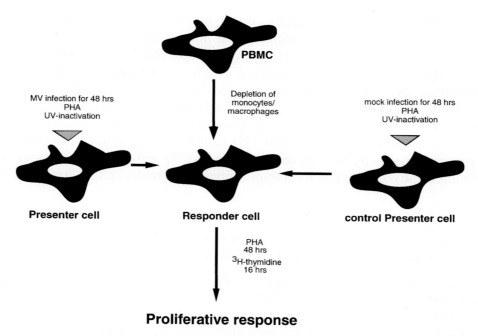

Proliferative response

Fig. 3. In vitro assay to study MV-induced immunosuppression. Mitogen-induced proliferation of responder cells (peripheral blood lymphocytes isolated from healthy human donors) as determined after a 16 h 3H-thymidine labeling period is investigated in the presence of MV-infected (or mock-infected) cells (presenter cells or control presenter cells, respectively). To abolish their proliferation and release of infectious MV, the presenter cells are UV-inactivated prior to cocultivation with the responder cells

was also observed using UV-inactivated MV, but not UV-inactivated vesicular stomatitis virus (VSV). A recombinant MV genetically modified to express the VSV glycoprotein G instead of the MV glycoproteins F and H [47, 70] used either to infect PCs, or after UV-inactivation, failed to interfere with RC proliferation, whereas coexpression of MV F and H in transient transfection assays was found to be necessary and sufficient to induce proliferative arrest of the RCs [52] (Fig. 4).

Although PC-induced unresponsiveness of RCs to mitogenic stimulation was long lasting and apparent even after four days, no evidence for the induction of apoptosis or cell loss by this interaction was found either in mitogen-stimulated PBLs or lymphocytic cell lines [68]. Rather, mitogen-stimulated upregulation of early activation markers including the IL-2R subunits as well as the release of cytokines such as IFN-γ or IL-10 were unimpaired for RCs after coculture with MV-infected PCs. Although levels of IL-2 released from these RCs were about 50% reduced as compared to mock-contacted controls, the anergic state was not reverted by IL-2 supplementation. MV glycoprotein-contact induced anergy was found associated with a strong cessation of cell cycling after mitogenic stimulation with an accumulation of RCs in the G1 phase of the cell cycle [68].

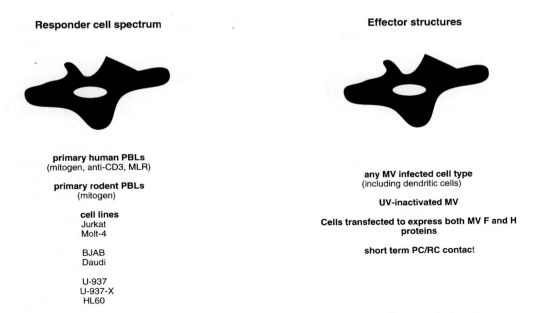

Responder cell spectrum

primary human PBLs
(mitogen, anti-CD3, MLR)

primary rodent PBLs
(mitogen)

cell lines
Jurkat
Molt-4

BJAB
Daudi

U-937
U-937-X
HL60

Effector structures

any MV infected cell type
(including dendritic cells)

UV-inactivated MV

Cells transfected to express both MV F and H proteins

short term PC/RC contact

Fig. 4. The expression of the MV glycoproteins is necessary and sufficent to induce immuno-suppression in vitro. A short membrane contact with the MV glycoproteins (expressed on MV-infected cells, on MV particles or cells transfected to express MV F and H proteins) is necessary and sufficient to induce unresponsiveness to mitogenic, allogenic (mixed leucocyte reaction, *MLR*), and anti-CD3-stimulation of primary human peripheral blood lymphocytes and mitogen stimulated rodent spleen cells. Moreover, proliferation of lymphocytic (T and B cell lines) and monocytic cell lines (U-937-X is a CD46-negative subclone of U-937 cells) is also inhibited in the presence of MV-infected presenter cells

MV interactions with dendritic cells

Dendritic cells are antigen-presenting cells specialised for the initiation of primary immune responses [72]. At different stages of their development they display a different functional repertoire. Immature dendritic cells are very effective in processing native protein antigens for the MHC class II restricted pathway. Mature dendritic cells are less able to capture new proteins for presentation but are much better at stimulating resting CD4+ and CD8+ T cells to grow and differentiate. Since dendritic cells are situated within mucous membranes, major sites of entry for many viruses, they are presumably also involved in controlling viral infection.

As professional antigen-presenting cells with numerous interactions with both naive and experienced T cells dendritic cells are likely to play a central role in the initiation of primary and secondary immune activation. For the very same reason, however, they could also efficiently contribute to MV-induced immuno-suppression if their interaction with MV would interfere with their allostimula-tory properties. Both immature precursor and mature dendritic cells were found to be highly susceptible to infection with MV in tissue culture as revealed by high accumulation levels of both cytoplasmic and surface MV proteins [23, 25, 69]. Similar to that in primary monocyte/macrophage cultures, virus release from infected dendritic cells was inefficient, suggesting that virus transmission may occur primarily in a cell-associated manner. Interestingly, the infection of both immature and mature dendritic cells with a MV wildtype strain proceeded significantly faster than that observed with the vaccine strain [69]. No obvious effects of MV infection on the expression of functionally important surface markers were found for mature dendritic cells. For immature dendritic cells, a rapid maturation was observed after MV infection as revealed by the upregulation of HLA-DR, CD40, CD83 and two important costimulatory molecules, CD80 and CD86 concommittant with the expression of viral glycoproteins. Conflicting results have been obtained regarding the release of IL-12 from dendritic cell cultures after MV-infection; whereas one study described a significant inhibition of CD40L-induced IL-12 release [23], MV-infection did not impair buth rather stimulated LPS- or SACS-induced IL-12 synthesis in our system [69]. As apparent from both studies, however, MV-infected dendritic cells largely failed to induce an allostimulatory response when assayed in a mixed leucocyte reaction, in spite of their highly activated phenotype. Moreover, even mitogen-dependent proliferation of PBLs was impaired in the presence of MV-infected dendritic cells (both with and without UV-inactivation). As also observed for functional maturation, wildtype MV-infected dendritic cells were stronger suppressors than those infected with the vaccine strain. Similar as outlined above for other PCs, no inhibitory effect was observed when a recombinant MV expressing the VSV G protein instead of the MV glycoproteins F and H was used for infection of the dendritic cells (Klagge et al., in prep.). These data indicate that even with mature dendritic cells expressing the full repertoire of costimulatory molecules the negative signal provided by the interaction with the MV F/H complex is dominant and efficiently impairs allogeneic and mitogen-dependent proliferation of RCs. Since dendritic cells functionally matured after MV-infection would be expected to home to the T cell rich area of the lymph nodes, these cells may play a central role for the induction of T cell anergy and, thus, measles associated immunosuppression.

Experimental infection of cotton rats to study MV-induced immunosuppression

Pathogenic aspects of MV immunosuppression have been difficult to study due to the lack of suitable animal modelds permissive for intranasal infection (see also above). Cotton rats (Sigmodon hispidus) are the only small animal model for the study of MV infection of lung and respiratory epithelium [77]. After intranasal infection of inbred cotton rats, MV replicates in lung tissue to peak titers after 4 to 5 days and subsides beginning on day 10. Histologically, a pneumonia with interstitial infiltration is seen. After exposure to MV, cotton rats mount a MV specific immune response and are immune against reinfection. Mitogen dependent proliferation of spleen cells from infected animals is inhibited, starting on day 3 up to day 7 with the strongest inhibition on day 4. Immune suppression is directly correlated with titers of MV recovered from lung tissue [43]. Similar to what has been described in tissue culture for the human system, mitogen unresponsiveness can also be induced using UV-inactivated MV, not however with the recombinant deficient for MV glycoprotein expression. Moreover, transfer of human fibroblasts doubly transfected with expression constructs for MV F and H (293-(F+H) cells) but not that of singly positive (293-F, or 293-H) or a mixture of both (293-F+293-H) was sufficient to induce spleen cell anergy. Thus, the expression of the viral glycoproteins on the surface of a relatively small number of cells was found necessary and sufficient for the induction of MV-induced immunosuppression in vitro and in vivo in an experimental animal model.

Conclusions and perspectives

Although we have learned much about the different strategies MV may use to establish a state that allows intracellular replication and escapes yet detection by immunological control during persistence, important questions have not been answered. These include whether persistence of MV is always a pathologic event or whether it is always established but needs to be triggered to cause pathology. If so, what would trigger resumption of active replication or development of disease? It is still unclear by which route, how frequently, and when during acute infection MV reaches the CNS, the only site at which a persistent MV infection in humans has been confirmed. Beyond any doubt, the receptor used for entry into neural cells is a main determinant of MV tropism in the CNS. It will be interesting to see to what extent CD46, which is expressed to very low levels on CNS cells, plays in this context or if alternative receptors, such as that involved in the attachment of wildtype MV isolates to lymphocytes, govern this primary interaction.

Obviously, the interaction with MV glycoproteins is not only important for viral entry but also for modulation of host immune responses. It remains for future analyses to reveal which domains within the MV glycoproteins are the actual effector structures for the induction of T cell anergy and which receptors they recognise on the surface of these cells. It will be of crucial importance to delineate which intracellular signalling events are activated or repressed by this interacation in order to develop strategies for therapeutic intervention.

Acknowledgements

The authors thank the Deutsche Forschungsgemeinschaft, the Robert Pfleger Stiftung, the Bundesministerium für Bildung und Forschung and the Humboldt foundation for financial support of their laboratory work.

References

1. Arneborn P, Biberfeld G (1983) T-Lymphocyte subpopulations in relation to immunosuppression in measles and varicella. Infect Immun 39: 29–37
2. Auwaerter P, Kaneshima H, McCune JM, Wiegand G, Griffin DE (1996) measles virus infection of thymic epithelium in the scid-hu mouse leads to thymocyte apoptosis. J Virol 70: 3 734–3 740
3. Baczko K, Pardowitz J, Rima BK, ter Meulen V (1992) Constant and variable regions of measles virus proteins encoded by the nucleocapsid and phosphoprotein genes derived from lytic and persistent viruses. Virology 190: 469–474

4. Baczko K, Lampe J, Liebert UG, Brinckmann U, ter Meulen V, Pardowitz J, Budka H, Cosby SL, Isserte S, Rima BK (1993) Clonal expansion of hypermutated measles virus in a SSPE brain. Virology 197: 188–195

5. Baczko K, Liebert UG, Billeter MA, Cattaneo R, Budka H, ter Meulen V (1986) Expression of defective measles virus genes in brain tissues of patients with subacute sclerosing panencephalitis. J Virol 59: 472–478

6. Barrett PN, Koschel K, Carter M, ter Meulen V (1985) Effect of measles virus antibodies on a measles SSPE virus persistently infected C6 rat-glioma cell line. J Gen Virol 66: 1 411–1 421

7. Bartz R, Brinckmann UG, Dunster LM, Rima BK, ter Meulen V, Schneider-Schaulies J (1996) Mapping amino acids of the measles virus haemagglutinin responsible for receptor (CD46) downregulation. Virology 224: 334–337

8. Bartz R, Firsching R, Rima BK, ter Meulen V, Schneider-Schaulies J (1998) Differential receptor usage by measles virus strains. J Gen Virol 79: 1 015–1 025

9. Billeter MA, Cattaneo R, Spielhofer P, Kaelin K, Huber M, Schmid A, Baczko K, ter Meulen V (1994) Generation and properties of measles virus mutations typically associated with subacute sclerosing panencephalitis. Ann NY Acad Sci 724: 367–377

10. Borrow P, Oldstone MBA (1995) Measles virus – mononuclear cell interactions. Curr Topics Microbiol Immunol 191: 85–100

11. Buchholz CJ, Gerlier D, Hu A, Cathomen T, Liszewski MK, Atkinson JP, Cattaneo R (1995) Selective expression of a subset of measles virus receptor competent CD46 isoforms in human brain. Virology 217: 349–355

12. Cattaneo R, Rebmann G, Schmid A, Baczko K, ter Meulen V, Billeter MA (1987) Altered transcription of a defective measles virus genome derived from a diseased human brain. EMBO J 6: 681–687

13. Cattaneo R, Rebmann G, Baczko K, ter Meulen V, Billeter MA (1987) Altered ratios of measles virus transcripts in diseased human brains. Virology 160: 523–526

14. Cattaneo R, Rose JK (1993) Cell fusion by the envelope glycoproteins of persistent measles viruses which caused lethal human brain diseases. J Virol 67: 1 493–1 502

15. Cattaneo R, Schmid A, Eschle D, Baczko K, ter Meulen V, Billeter MA (1988) Biased hypermutation and other genetic changes in defective measles viruses in human brain infections. Cell 55: 255–265

16. Doerig RE, Marcil A, Chopra A, Richardson CD (1993) The human CD46 Molecule is a receptor for measles virus (Edmonston Strain). Cell 75: 295–305

17. Dunster LM, Schneider-Schaulies J, Löffler S, Lankes W, Schwarz-Albiez R, Lottspeich F, ter Meulen V (1994) Moesin: A cell membrane protein linked with suceptibility to measles virus infection. Virology 198: 265–274

18. Dunster LM, Schneider-Schaulies J, Dehoff MH, Holers VM, Schwartz-Albiez R, ter Meulen V (1995) Moesin, and not the murine functional homologue (crry/p65) of human membrane cofactor protein (CD46) is involved in the entry of measles virus (strain Edmonston) into susceptible murine cell lines. J Gen Virol 76: 2 085–2 089

19. Ecker A, ter Meulen V, Baczko K, Schneider-Schaulies S (1995) MV-specific dsRNAs are targets for unwinding modifying activity in vitro. J Neurovirol 1: 92–100

20. Esolen LM, Ward BJ, Moench TR, Griffin DE (1993) Infection of monocytes during measles. J Infect Dis 168: 47–52

21. Esolen LE, Park SW, Hardwick JM, Griffin DE (1995) Apoptosis as a cause of death in measles virus-infected cells. J Virol 69: 3 955–3 958

22. Galama JMD, Ubels-Postma J, Vos A, Lucas CJ (1980) Measles virus inhibits acquisition of lymphocyte functions but not established effector functions. Cell Immunol 50: 405–415

23. Fugier-Vivier I, Servet-Delprat C, Rivailler P, Riossan MC, Liu YJ, Rabourdin-Combe C (1997) Measles virus suprresses cell-mediated immunity by interfering with the survival and function of dendritic and T cells. J Exp Med 186: 813–823

24. Griffin DE (1995) Immune responses during measles virus infecation. Curr Topics Microbiol Immunol 191: 117–134

25. Grosjean I, Caux C, Bella C, Berger D, Wild TF, Banchereau J, Kaiserlian D (1997) Measles virus infects human dendritic cells and blocks their allostimulatory properties for CD4+ T cells. J Exp Med 186: 801–812

26. Horikami S, Moyer SA (1995) Double stranded RNA adenosine deaminase activity during measles virus infection. Virus Res 36: 87–96

27. Horikami S, Moyer SA (1995) Structure, transcription and replication of measles virus. Curr Topics Microbiol Immunol 191: 35–50

28. Hsu EC, Sarangi F, Iorio C, Sidhu MS, Udem SA, Dillehay DL, Xu W, Rota PA, Bellini WJ, Richardson CD (1998) A single amino acid change in the haemagglutinin protein of measles virus determines its ability to bind CD46 and reveals another receptor on marmoset B cells. J Virol 72: 2 905–2 916

29. Johnstone RW, Loveland B, Mckenzie IFC (1993) Identification and quantification of complement regulator CD46 on normal human tissues. Immunology 79: 341–347

30. Karp CL, Wysocka M, Wahl LM, Ahearn JM, Cuomo PJ, Sherry B, Trinchieri G, Griffin DE (1996) Mechanism of suppression of cell-mediated immunity by measles virus. Science 273: 228–231

31. Katz M (1995) Clinical spectrum of measles. Current Top Microbiol Immunol 191: 1–13

32. Krantic S, Giminez C, Rabourdin-Combe C (1995) Cell-to-cell contact via measles virus haemagglutinin-CD46 interaction triggers CD46 downregulation. J Gen Virol 76: 2 793–2 800

33. Komase K, Rima BK, Pardowitz J, Kunz C, Billeter MA, ter Meulen V, Baczko K (1995) A comparison of nucleotide sequences of measles virus L genes derived from wild type viruses and SSPE brain tissues. Virology 208: 795–799

34. Kraus E, Schneider-Schaulies S, Miyasaka M, Tamatani T, Sedgwick J (1992) Augmentation of major histocompatibility complex class I and ICAM-1 expression on glial cells following measles virus infection: evidence for the role of type-1 interferon. Eur J Immunol 22: 175–182

35. Lecouturier V, Fayolle J, Caballero M, Carabana J, Celma ML, Fernandez-Munoz R, Wild TF, Buckland R (1996) Identification of two acids in the haemagglutinin glycoprotein of measles virus (MV) that govern hemadsorption, Hela cell fusion and CD46 downregulation: phenotypic markers that differentiate vaccine and wild-type MV strains. J Virol 70: 4 200–4 204

36. Liebert UG, Baczko K, Budka H, ter Meulen V (1986) Restricted expression of measles virus proteins in brains from cases of subacute sclerosing panencephalitis. J Gen Virol 67: 2 435–2 444

37. Liebert UG, Schneider-Schaulies S, Baczko K, ter Meulen V (1990) Antibody-induced restriction of viral gene expression in measles encephalitis in rats. J Virol 64: 706–713

38. McChesney MB, Kehrl JH, Valsamakis A, Fauci AS, Oldstone MBA (1987) Measles virus infection of B lymphocytes permits cellular activation but blocks progression through the cell cycle. J Virol 61: 3 441–3 447

39. ter Meulen V, Stephenson JR, Kreth HW (1983) Subacute sclerosing panencephalitis. In: Fraenkel-Conrat H, Wagner RR (eds) Comprehensive virology. Elsevier, Amsterdam, pp 105–159

40. Nakayama T, Mori T, Yamaguchi S, Sonoda S, Asamura A, Yamashita R, Takeuchi Y, Urano T (1995) Detection of measles virus genome directly from clinical samples by

reverse transcriptase-polymerase chain reaction and genetic variability. Virus Res 35: 1–16

41. Naniche D, Varior-Krishnan G, Cervoni F, Wild TF, Rossi B, Rabourdin-Combe C, Gerlier D (1993) Human membrane cofactor protein (CD46) acts a cellular receptor for measles virus. J Virol 67: 6 025–6 032

42. Niewiesk S, Schneider-Schaulies J, Ohnimus H, Jassoy C, Schneider-Schaulies S, Diamond L, Logan JS, ter Meulen V (1997) CD46 expression does not overcome the intracellular block of measles virus replication in transgenic rats. J Virol 71: 7 969–7 973

43. Niewiesk S, Eisenhuth I, Fooks A, Clegg JC, Schnorr JJ, Schneider-Schaulies S, ter Meulen V (1997) Measles virus induced immune suppression in the cotton rat (Sigmodon hispidus) model depends on viral glycoproteins. J Virol 71: 7 214–7 219

44. Nussbaum O, Broder CC, Moss B, Stern B, Rozenblatt S, Berger EA (1995) Functional and structural interactions between measles virus haemagglutinin and CD46. J Virol 69: 3 341–3 349

45. Ohuchi M, Ohuchi R, Mifune K, Ishihara T, Ogawa T (1987) Characterization of the measles virus isolated from the brain of a patient with immunosuppressive measles encephalitis. J Infect Dis 156: 436–441

46. Patterson JB, Samuel CE (1995) Expression and regulation by interferon of a double stranded-RNA-specific adenosine desaminase from human cells: evidence for two forms of the deaminase. Mol Cell Biol 15: 5 376–5 388

47. Radecke F, Spielhofer P, Schneider H, Kaelin K, Huber M, Dötsch C, Christiansen G, Billeter MA (1995) Rescue of measles virus from cloned cDNA. EMBO J 14: 5 773–5 784

48. Rager-Zisman B, Egan JE, Kress Y, Bloom BR (1984) Isolation of cold-sensitive mutants of measles virus from persistently infected murine neuroblastoma cells. J Virol 51: 845–855

49. Rataul SM, Hirano A, Wong TC (1992) Irreversible modification of measles virus RNA in vitro by nuclear RNA-unwinding activity in human neuroblastoma cells. J Virol 66: 1 769–1 773

50. Rima BK, Earle JAP, Baczko K, Rota PA, Bellini WJ (1995) Measles virus strain variations. Curr Topics Microbiol Immunol 191: 65–83

51. Sanchez-Lanier M, Guerlin P, Mclaren LC, Bankhurst AD (1988) Measles virus induced suppression of lymphocyte proliferation. Cell Immunol 116: 367–381

52. Schlender J, Schnorr JJ, Spielhofer P, Cathomen T, Cattaneo R, Billeter MA, ter Meulen V, Schneider-Schaulies S (1996) Interaction of measles virus glycoproteins with the surface of uninfected peripheral blood lymphocytes induces immunosuppression in vitro. Proc Natl Acad Sci USK 93: 13 194–13 199

53. Schmid A, Cattaneo R, Billeter MA (1987) A procedure for selective full length cDNA cloning of specific RNA species. Nucleic Acids Res 15: 3 987–3 996

54. Schmid A, Spielhofer P, Cattaneo R, Baczko K, ter Meulen V, Billeter MA (1992) Subacute sclerosing panencephalitis is typically characterized by alteration in the fusion protein cytoplasmic domain of persisting measles virus. Virology 188: 910–915

55. Schneider-Schaulies J, Dunster LM, Schwartz-Albiez R, Krohne G, ter Meulen V (1995) Physical association of moesin and CD46 as a receptor complex for measles virus. J Virol 69: 2 248–2 256

56. Schneider-Schaulies J, Schnorr JJ, Brinckmann U, Dunster LM, Baczko K, Schneider-Schaulies S, ter Meulen V (1995) Receptor usage and differential downregulation of CD46 by measles virus wild type and vaccine strains. Proc Natl Acad Sci USA 92: 3 943–3 947

57. Schneider-Schaulies J, Dunster LM, Kobune F, Rima BK, ter Meulen V (1995) Differential downregulation of CD46 by measles virus strains. J Virol 69: 7 257–7 259

58. Schneider-Schaulies J, Schnorr JJ, Schlender J, Dunster LM, Schneider-Schaulies S, ter Meulen V (1996) Receptor (CD46) modulation and complement mediated lysis of uninfected cells after contact with measles virus infected cells. J Virol 70: 255–263

59. Schneider-Schaulies S, Kreth HW, Hofmann G, Billeter M, ter Meulen V (1991) Expression of measles virus RNA in peripheral blood mononuclear cells of patients with measles, SSPE, and autommue diseases. Virology 182: 703–711

60. Schneider-Schaulies S, ter Meulen V (1992) Molecular aspcts of measles virus induced central nervous system diseases. In Ross RP (ed) Molecular neurovirology. Humana Press, Clifton, pp 419–449

61. Schneider-Schaulies S, Schneider-Schaulies J, Bayer M, Löffler S, ter Meulen V (1993) Spontaneous and differentiation dependent regulation of measles virus gene expression in human glial cells. J Virol 67: 3 375–3 383

62. Schneider-Schaulies S, Liebert UG, Baczko K, ter Meulen V (1990) Restricted expression of measles virus in primary rat astroglial cells. Virology 177: 802–806

63. Schneider-Schaulies S, Liebert UG, Baczko K, Cattaneo R, Billeter M, ter Meulen V (1989) Restriction of measles virus gene expression in acute and subacute encephalitis of Lewis rats. Virology 171: 525–534

64. Schneider-Schaulies S, Liebert UG, Segev Y, Rager-Zisman B, Wolfson M, ter Meulen V (1992) Antibody-dependent transcriptional regulation of measles virus in persistently infected neural cells. J Virol 66: 5 534–5 541

65. Schneider-Schaulies S, Schuster A, Schneider-Schaulies J, Bayer M, Pavlovic J, ter Meulen V (1994) Cell type specific MxA-mediated inhibition of measles virus transcription in human brain cells. J Virol 68: 6 910–6 917

66. Schnorr JJ, Schneider-Schaulies S, Simon-Jödicke A, Pavlovic J, Horisberger MA, ter Meulen V (1993) MxA dependent inhibition of measles virus glycoprotein synthesis in a stably transfected human monocytic cell line. J Virol 67: 4 760–4 768

67. Schnorr JJ, Dunster LM, Nanan R, Schneider-Schaulies J, Schneider-Schaulies S, ter Meulen V (1995) Measles virus induced downregulation of CD46 is associated with enhanced sensitivity to complement mediated lysis of infected cells. Eur J Immunol 25: 976–984

68. Schnorr JJ, Seufert M, Schlender J, Borst J, Johnston ICD, ter Meulen V, Schneider-Schaulies S (1997) Cell cycle arrest rather than apoptosis is associated with measles virus contact-mediated immunosuppression in vitro. J Gen Virol 78: 3 217–3 226

69. Schnorr JJ, Xanthakos S, Keikavoussi P, Kämpgen E, ter Meulen V, Schneider-Schaulies S (1997) Induction of maturation of human blood dendritic cell precursors by measles virus is associated with immunosuppression. Proc Natl Acad Sci USA 94: 5 326–5 331

70. Spielhofer P, Bächi T, Fehr T, Christiansen G, Cattaneo R, Kaelin K, Billeter MA, Naim HY (1998) Chimeric measles viruses with a foreign envelope. J Virol 72: 2 150–2 159

71. Schuster A, Johnston ICD, Das T, Banerjee AK, Pavlovic J, Schneider-Schaulies S (1996) MxA expression in human brain cells is associated with hyperphosphorylation of P protein of VSV. Virology 220: 241–245

72. Steinman RM (1991) The dendritic cell system and its role in immunogenicity. Ann Rev Immunol 9: 271–196

73. Vanchiere JA, Bellini WJ, Moyer SA (1995) Hypermutation of the phosphoprotein and altered mRNA editing in the hamster neurotropic strain of measles. Virology 207: 555–5 561

74. Watanabe M, Hirano A, Stenglein S, Nelson J, Thomas G, Wong TC (1995) Engineered

serine protease inhibitor prevent furin-catalysed activation of the fusion glyprotein and production of infectious measles virus. J Virol 69: 3 206–3 210

75. Wild TF, Malvoisin E, Buckland R (1991) Measles virus: both the haemagglutinin and the fusion glycoproteins are required for fusion. J Gen Virol 72: 439–442

76. Wong TC, Ayata M, Ueda S, Hirano A (1991) Role of baised hypermutation in evolution of subacute sclerosing panencephalitis virus from progenitor acute measles virus. J Virol 65: 2 191–2 199

77. Wyde PR, Ambrosi MW, Voss TG, Meyer HL, Gilbert BF (1992) Measles virus replication in lungs of hispid cotton rats after intranasal inoculation. Proc Soc Exp Biol Med 24: 80–87

78. Yanagi Y, Cubitt BA, Oldstone MBA (1992) Measles virus inhibits mitogen-induced T cell proliferation. Virology 187: 280–289

79. Yoshikawa Y, Yamanouchi K (1984) Effects of papaverine treatment on replication of measles virus in human neural and non-neural cells. J Virol 50: 489–495

Authors' address: Dr. S. Schneider-Schaulies, Institute of Virology and Immunobiology, University of Würzburg, Versbacher Str. 7, D-97078 Würzburg, Germany.

The glycoproteins of Marburg and Ebola virus and their potential roles in pathogenesis

H. Feldmann, V. E. Volchkov, V. A. Volchkova, and **H.-D. Klenk**

Institut für Virologie, Philipps-Universität Marburg, Marburg, Germany

Summary. Filoviruses cause systemic infections that can lead to severe hemorrhagic fever in human and non-human primates. The primary target of the virus appears to be the mononuclear phagocytic system. As the virus spreads through the organism, the spectrum of target cells increases to include endothelial cells, fibroblasts, hepatocytes, and many other cells. There is evidence that the filovirus glycoprotein plays an important role in cell tropism, spread of infection, and pathogenicity. Biosynthesis of the glycoprotein forming the spikes on the virion surface involves cleavage by the host cell protease furin into two disulfide linked subunits GP_1 and GP_2. GP_1 is also shed in soluble form from infected cells. Different strains of Ebola virus show variations in the cleavability of the glycoprotein, that may account for differences in pathogenicity, as has been observed with influenza viruses and paramyxoviruses. Expression of the spike glycoprotein of Ebola virus, but not of Marburg virus, requires transcriptional editing. Unedited GP mRNA yields the nonstructural glycoprotein sGP, which is secreted extensively from infected cells. Whether the soluble glycoproteins GP_1 and sGP interfere with the humoral immune response and other defense mechanisms remains to be determined.

Introduction

Filoviruses cause fulminant hemorrhagic fever in humans and non-human primates, killing up to 90% of the infected patients. Since the discovery of Marburg virus in 1967 and the emergence of Ebola virus, its better known cousin, a few years later, these infections have therefore been a matter of high public and scientific concern. Although it is clear from the recorded history of filovirus outbreaks that all of them have so far been self-limiting and that the total number of human infections hitherto documented scarcely exceeds a thousand cases, Ebola virus by now ranges among the most ill-famed human viruses. For a long time, research on filoviruses has been impeded by their high pathogenicity, but with the advent of recombinant DNA technology our knowledge of the genome structures and the replication strategies of these agents has significantly increased.

Pathophysiology of filovirus infections

The pathophysiological changes that make filovirus infections so devastating are just beginning to be unraveled. Pathogenesis in fatal infections in human and non-human primates is similar, suggesting the primate system as a model for studying filovirus hemorrhagic fever [9, 17, 28, 29, 38]. Clinical and biochemical findings support the anatomical observations of extensive liver involvement, renal damage, changes in vascular permeability including endothelial damage, and activation of the clotting cascade. The visceral organ necrosis is a consequence of virus replication in parenchymal cells. However, no organ, not even liver, shows sufficient damage to account for death. The role of disseminated intravascular coagulation (DIC) in pathogenesis is still controversial, since a laboratory confirmation of DIC in human infections has never been demonstrated. In non-human primates the intrinsic clotting pathway is most affected whereas the extrinsic pathway is spared. The consequence is a DIC in final stages of the infection when parenchymal necrosis is extensive.

Fluid distribution problems and platelet abnormalities are dominant clinical manifestations reflecting damage of endothelial cells and decrease of platelets. Post mortem there is little monocyte/macrophage infiltration in sites of parenchymal necrosis, suggesting that a dysfunction of white blood cells, such as macrophages, also occurs. Morphological studies on monkeys infected with the Reston subtype of Ebola virus from the 1989 epizootic [19] and monkeys experimentally infected with the Zaire subtype [29] showed that monocytes/macrophages and fibroblasts may be the preferred sits of virus replication in early stages, whereas other cell types may become involved as the disease progresses. Human monocytes/macrophages in culture are also sensitive to infection, resulting in massive production of infectious virus and cell lysis [12]. Although the studies on infected non-human primates did not identify endothelial cells as sites of massive virus replication, in vitro studies and post mortem observations of human cases clearly demonstrated that endothelial cells of human origin are suitable targets for virus replication [35, 53]. Here infection leads to cell lysis, indicating that damage of endothelial cells may be an important pathophysiological parameter during infection.

In addition to evidence for direct vascular involvement in infected hosts, the role of active mediator molecules in the pathogenesis of the disorders must be discussed. It has been demonstrated that supernatants of filovirus-infected monocyte/macrophage cultures are capable of increasing paraendothelial permeability in an in vitro model [10]. Examination for mediators in those supernatants revealed increased levels of secreted TNF-α, the prototype cytokine of macrophages. These data support the concept of a mediator-induced vascular instability and, thus, increased permeability as a key mechanism for the development of the shock syndrome seen in severe and fatal cases. Thus, the syndrome may be comparable to shock in response to various endogenous and exogenous mediators [36]. The bleeding tendency could be due to endothelial damage caused directly by virus replication as well as indirectly by cytokine-mediated processes. The onset of the

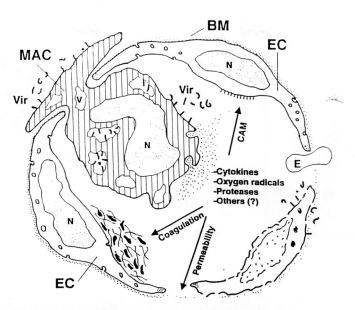

Fig. 1. Schematic drawing illustrating the possible role of macrophages and endothelial cells in the development of hemorrhagic fever caused by filoviruses. *EC* Endothelial cell; *MAC* macrophage; *Vir* virus particles; *CAM* cell adhesion molecule; *E* erythrocyte; *BM* basement membrane; *N* nucleus; *V* vacuole. Taken from [10]

bleeding tendency is supported by the loss of the integrity of the endothelium, as demonstrated in tissue and organ culture [35] as well as in infected animals [17] and seems to occur later in infection. The bleeding tendency may be reinforced by a decrease of the bloodstream as a common consequence of shock. The combination of viral replication in endothelial cells and virus-induced cytokine release from monocytes/macrophages may also promote a distinct proinflammatory endothelial phenotype that then triggers the coagulation cascade. A model summarizing these pathophysiological events is illustrated in Fig. 1.

Genome and genome products of filoviruses

Marburg virus and Ebola virus belong to the family *Filoviridae*, order *Mononegavirales*. They have non-segmented negative-stranded RNA genomes that encode the seven structural proteins in the order nucleoprotein (NP), virion structural protein (VP) 35, VP40, glycoprotein (GP), VP30, VP24, and RNA-dependent RNA polymerase (L) [13, 31, 44] (Fig. 2). In general, filoviral genes are transcribed into monocistronic subgenomic RNA species (mRNA) [13, 30]. In contrast to all other filoviral genes, including the GP gene of Marburg virus [49], the organization and transcription of the fourth gene (GP) of Ebola virus is unusual, involving transcriptional editing needed to express the envelope glycoprotein (Fig. 3). A nonstructural small glycoprotein (sGP) is synthesized from the unedited GP mRNA, which is extensively secreted from infected cells [33, 42]. It has been reported that by binding to neutrophils sGP may interfere with the

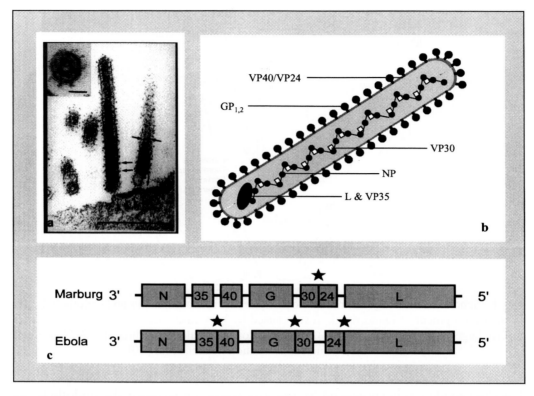

Fig. 2. The structure of filoviruses. **a** Electron micrograph of Marburg virus particles budding from the surface of human endothelial cells 3 days after infection. Particles consist of a nucleocapsid surrounded by a membrane in which spikes are inserted (arrows). The nucleocapsid contains a central channel (inset). The plasma membrane of infected cells is often thickened at locations where budding occurs (arrowheads) (bar 0.5 μm; inset bar=50 nm). **b** Schematic view of a virion. The RNA genome is associated with 4 viral proteins: the viral polymerase (*L*), the nucleoprotein (*NP*), and proteins VP35 and VP30. VP40 and VP24 are matrix proteins. The spikes are formed by trimeric $GP_{1,2}$ complexes. **c** The gene order on the non-segmented negativ-strand RNA genome of Marburg and Ebola viruses. Overlapping genes are indicated by asterisks. Taken from [14]

activation of these cells and thereby may paralyze the inflammatory defense of the host [52]. A third mRNA of the GP gene encoding another small glycoprotein (ssGP) has also been identified [42]. The envelope glycoprotein GP of Marburg and Ebola virus is a trimeric type I transmembrane glycoprotein [11, 32]. The middle region of GP is variable, extremely hydrophilic, and carries the bulk of the glycosylation sites for N- and O-glycans that account for approximately one third of the molecular weight [4, 11, 20, 44]. Experimental data on GP function are limited. However, the fact that GP is the only surface protein of virions and that it mediates infection by vesicular stomatitis virus [40] and retrovirus pseudotypes [51, 52] suggest a function in receptor binding and fusion with cellular membranes. There is also evidence that Marburg virus infects hepatocytes by binding to the asialoglycoprotein receptor of these cells [3]. Maturation of Ebola virus GP

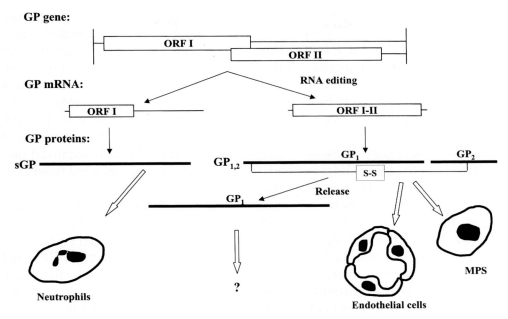

Fig. 3. Biosynthesis of different forms of the Ebola virus glycoprotein and their target cells. The Ebola virus surface glycoprotein GP is encoded in two overlapping reading frames (ORF I and II), and expression of GP occurs through transcriptional editing. ORF I encodes for a secreted small glycoprotein (sGP) that is expressed from unedited transcripts [33, 42]. The amino-terminal 295 amino acids of sGP are identical with GP, but the last 69 ones are different. Mature GP ($GP_{1,2}$) consists of the disulfide-linked (S-S) subunits GP_1 and GP_2 [45]. Significant amounts of GP_1 are released from expressing cells [46]. sGP binds to neutrophils and appears to inhibit early activation of these cells [52], whereas GP mediates binding of Ebola virus to endothelial cells and presumably to other cells susceptible to infection. The binding specificity of GP_1 is not known yet. Taken from [26]

involves posttranslational cleavage of a precursor into the disulfide-linked fragments GP_1 and GP_2 [45]. GP_1 is not only present in virion spikes, but is also found in soluble form after shedding from the surface of infected cells. Furthermore, membrane vesicles containing the complete envelope glycoprotein are released from infected cells, an action that may be partly responsible for immune modulatory effects associated with Ebola virus infection [46]. Thus, unlike Marburg virus glycoprotein, Ebola virus glycoprotein shows a much higher degree of polymorphism, and there is evidence that the different products of the GP gene have different functions in the infected host (Fig. 3).

Proteolytic processing of the filovirus glycoprotein as a potential determinant of pathogenicity

GP is cleaved by furin [45]. This is indicated by the observation that cleavage did not occur when GP was expressed in the furin-defective LoVo cell line, but that it was restored in these cells by vector-expressed furin. The finding that cleavage was inhibited by a sequence-specific peptidyl chloromethylketone or by mutation of the cleavage site supports this concept. Furin belongs to the proprotein convertases, a family of subtilisin-like eukaryotic endoproteases that includes also PC1/PC3, PC2, PC4, PACE4, PC5/PC6, and LPC/PC7 [37]. These enzymes are differentially expressed in cells and tissues, and they display similar but not identical specificity for basic motifs, such as R-X-K/R-R, at the cleavage site of their substrates. Furin appears to be expressed in most cells. It is a processing enzyme of the constitutive secretory pathway, as seems to be the case with PACE4, PC5/PC6, and LPC/PC7. The expression of PC1/PC3 and PC2 is restricted to the regulated secretory pathway of neuroendocrine cells. Furin is localized predominantly in the trans-Golgi network [27, 34], but it is also secreted from cells in a truncated form [41, 50]. Proprotein convertases activate numerous cellular proteins [2] and surface proteins of enveloped viruses. Furin appears to be the key enzyme in virus activation [23], but PC5/PC6 [22] and LPC/PC7 [21] are also involved. Thus, LPC/PC7 may be responsible for cleavage of the HIV glycoprotein in the furin-deficient LoVo cells. The observation that Ebola virus GP is not cleaved in these cells is interesting in this context. It is also noteworthy that furin, although ubiquitous, is particularly apparent in hepatocytes and endothelial cells, which are both prime targets of Ebola virus [19, 53]. These observations stress the importance of furin as a processing enzyme of GP, but it remains to be seen in future studies if other proprotein convertases can substitute as cleaving enzymes.

Processing by protein convertases is an important control mechanism for the biological activity of viral surface proteins [23, 24]. Cleavage occurs often next to a protein domain involved in fusion, and it has long been known that in these cases proteolytic cleavage is necessary for fusion activity. Proteolytic cleavage is the first step in the activation of these fusion proteins and is followed by a conformational change resulting in the exposure of the fusion domain [5, 7, 48]. The conformational change may be triggered by low pH in endosomes, as is the case with influenza viruses [39], or by the interaction with a secondary receptor protein at the cell surface, as is the case with HIV [15]. We have so far not been able to demonstrate that cleavage of GP has an effect on fusion activity or on infectivity of Ebola virus. However, it is interesting to see that GP_2 contains a sequence of 16 uncharged and hydrophobic amino acids at a short distance (22 amino acids) from the cleavage site which bears some structural similarity to the fusion peptides of retroviruses and has therefore been thought to play a role in Ebola virus entry [18]. Furthermore, it appears that cleaved and uncleaved GP differ in folding, as indicated by their differential electrophoretic mobilities under non-reducing conditions. Finally, the central structural feature of the GP_2 ectodomain is a long triple-stranded coiled coil with three antiparallel helices packed at the surface of this trimer, as has been observed with other viral fusion proteins undergoing proteolytic activation (Fig. 4). These observations are compatible with the view that proteolytic cleavage is a priming mechanism that renders GP susceptible to the conformational change required for fusion.

Finally, it has to be pointed out that proteolytic activation of viral glycoproteins is an important determinant for pathogenicity. Cleavage by furin and other ubiquitous proprotein convertases has been shown to be responsible for

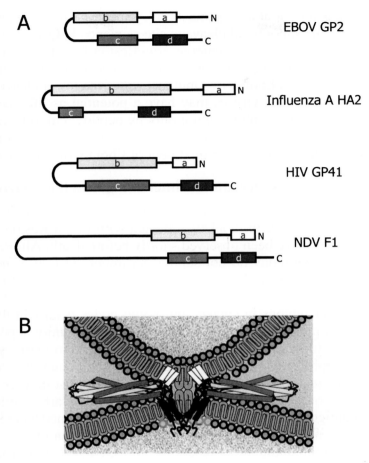

Fig. 4. The filovirus glycoprotein is a potential fusion protein. **A** Structural similarities between EBOV GP2 [47] and the transmembrane subunits HA2 of the influenza virus hemagglutinin [5, 8], gp41 of the HIV env protein [7, 48] and the F1 of the Newcastle disease virus fusion protein [6]. Four domains can be discriminated in the fusion active state: the fusion peptide (*a*), an amino-terminal helix (*b*), a carboxy-terminal helix (*c*), and the membrane anchor (*d*). The transmembrane proteins assemble to trimers in which the large amino-terminal helices form an interior, parallel coiled-coil with the smaller carboxy-terminal helices packing in an antiparallel fashion at the surface. The fusion peptide and the membrane anchor are therefore located at one end of the rod-like trimers. **B** Fusion model. The close proximity of the fusion peptide and the membrane anchor brings both membranes together and thereby promotes fusion [48]

systemic infection caused by highly pathogenic strains of avian influenza and Newcastle disease virus [25]. It is therefore tempting to speculate that cleavage by furin is also an important factor for the pantropism of Ebola virus and its rapid dissemination through the organism. Furthermore, variations at the clevage site of GP may account for differences in the pathogenicity of Ebola virus [45]. The pathogenic strains Zaire, Sudan, and Ivory Coast which have the canonical furin motif R-X-K/R-R at the cleavage site are highly susceptible to cleavage, whereas the Reston strain which appears to be apathogenic for humans and only moderately pathogenic for at least some monkey species [16] has reduced cleavability because of the suboptimal cleavage site sequence K-Q-K-R. That highly pathogenic variants may suddenly emerge from Reston-like strains by mutations restricted to the cleavage site is an intriguing hypothesis. On the other hand, it may be possible to obtain Ebola virus mutants with even lower cleavability than the Reston strain, and such viruses may have a potential as life vaccines. Because furin cleavage can be inhibited not only by peptidyl chloromethylketones as described here but also by less toxic components [1], inhibition of proteolytic cleavage may be a novel concept for treatment of Ebola virus infections.

Potential filovirus mechanisms interfering with host defense

Fatal filovirus infections usually end with high viremia and no evidence of an effective immune response. In monkeys infected with Ebola-Reston virus non-protective antibodies have been observed shortly before death. Altogether, however, the data available today do not support an important role of neutralizing antibodies in virus clearance. Because circulating monocytes/macrophages are primary target cells in filovirus infections and because the extensive disruption of the parafollicular regions in spleen and lymph nodes results in the destruction of the antigen-presenting dendritic cells, cellular immunity appears also to be affected during filoviral hemorrhagic fever. In addition to these cytolytic effects, the polymorphism of the Ebola virus glycoprotein suggests a number of other mechanisms regarding virus interference with the host defense. First, the immune reactivity of $GP_{1,2}$ may be modulated by its high carbohydrate content that may cover antigenic epitopes. Second, as already mentioned there is evidence that by binding to neutrophils sGP may block the activation of these cells and thereby interfere with inflammatory reactions. Third, GP_1 released from infected cells by shedding may have a decoy function by binding to GP-specific antibodies. Fourth, a presumably immunosuppressive domain has been identified on GP_2 [43]. Whether any of these mechanisms contributes to pathogenicity remains to be shown in future studies.

References

1. Anderson ED, Thomas L, Hayflick JS, Thomas G (1993) Inhibition of HIV-1 gp 160-dependent membrane fusion by a furin-directed alpha 1-antitrypsin variant. J Biol Chem 268: 24 887–24 891
2. Barr PJ (1991) Mammalian subtilisins: the long-sought dibasic processing endoproteases. Cell 66: 1–3
3. Becker S, Spiess M, Klenk H-D (1995) The asialoglycoprotein receptor is a potenital liver-specific receptor for Marburg virus. J Gen Virol 76: 393–399
4. Becker S, Klenk H-D, Mühlberger E (1996) Intracellular transport and processing of the Marburg virus surface protein in vertebrate and insect cells. Virology 225: 145–155
5. Bullough PA, Hughson FM, Skehel JJ, Wiley DC (1994) Structure of influenza haemagglutinin at the pH of membrane fusion. Nature 371: 37–43
6. Chambers P, Pringle CR, Easton AJ (1990) Heptad repeat sequences are located adjacent to hydrophobic regions in several types of virus fusion glycoproteins. Gen Virol 71: 3 075–3 080

7. Chan DC, Fass D, Berger JM, Kim PS (1997) Core structure of gp41 from the HIV envelope glycoprotein. Cell 89: 263–273

8. Carr CM, Kim PS (1993) A spring-loaded mechanism for the conformational change of influenza hemagglutinin. Cell 21: 823–832

9. Ellis DS, Bowen ETW, Simpson DIH (1978) Ebola virus: a comparison, at ultrastructural level, of the behaviour of the Sudan and Zaire strains in monkeys. Br J Exp Pathol 59: 584–593

10. Feldmann H, Klenk H-D (1996) Marburg and Ebola viruses. Adv Virus Res 47: 1–52

11. Feldmann H, Will C, Schikore M, Slenczka W, Klenk H-D (1991) Glycosylation and oligomerization of the spike protein of Marburg virus. Virology 182: 353–356

12. Feldmann H, Bugany H, Mahner F, Klenk H-D, Drenckhahn D, Schnittler H-J (1996) Filovirus-induced endothelial leakage triggered by infected monocytes/macrophages. J Virol 70: 2 208–2 214

13. Feldmann H, Mühlberger E, Randolf A, Will C, Kiley MP, Sanchez A, Klenk H-D (1992) Marburg virus, a filovirus: messenger RNAs, gene order, and regulatory elements of the replication cycle. Virus Res 24: 1–19

14. Feldmann H, Volchkov VE, Klenk H-D (1997) Filovirus Marburg et Ebola. Ann Inst Pasteur 8: 207–222

15. Feng Y, Broder CC, Kennedy PE, Berger EA (1996) HIV-1 entry cofactor: functional cDNA cloning of a seven-transmembrane, G protein-coupled receptor. Science 272: 872–877

16. Fisher-Hoch SP, Brammer TL, Trappier SG, Hutwagner LC, Farrar BB, Ruo SL, Brown BG, Hermann LM, Perez-Oronoz GI, Goldsmith CS, Hanes MA, McCormick JB (1992) Pathogenic potential of Filoviruses: role of geographic origin of primate host and virus strain. J Infect Dis 166: 753–763

17. Fisher-Hoch SP, Platt GS, Neild GH, Southee T, Baskerville A, Raymond RT, Lloyd G, Simpson DIH (1985) Pathophysiology of shock and hemorrhage in a fulminating viral infection (Ebola). J Infect Dis 152: 887–894

18. Gallaher WR (1996) Similar structural models of the transmembrane proteins of Ebola and avian sarcoma viruses. Cell 85: 477–478

19. Geisbert TW, Jahrling PB, Hanes MA, Zack PM (1992) Association of Ebola-related Reston virus particles and antigen with tissue lesions of monkeys imported to the United States. J Comp Path 106: 137–152

20. Geyer H, Will C, Feldmann H, Klenk H-D, Geyer R (1992) Carbohydrate structure of Marburg virus glycoprotein. Glycobiology 2: 299–312

21. Hallenberger S, Moulard M, Sordel M, Klenk H-D, Garten W (1997) The role of eukaryotic subtilisin-like endoproteases for the activation of human immunodeficiency virus glycoproteins in natural host cells. J Virol 71: 1 036–1 045

22. Horimoto T, Nakayama K, Smeekens SP, Kawaoka Y (1994) Proprotein-processing endoproteases PC6 and furin both activate hemagglutinin of virulent avian influenza viruses. J Virol 68: 6 074–6 078

23. Klenk H-D, Garten W (1994a) Activation cleavage of viral spike proteins by host proteases. In: Wimmer E (ed) Cellular receptors for animal viruses, Cold Spring Harbor Laboratory Press, New York, pp 241–280

24. Klenk H-D, Garten W (1994b) Host cell proteases controlling virus pathogenicity. Trends Microbiol 2: 39–43

25. Klenk H-D, Rott R (1988) The molecular biology of influenza virus pathogenicity. Adv Virus Res 34: 247–281

26. Klenk H-D, Volchkov VE, Feldmann H (1998) Two strings to the bow of Ebola virus. Nature Med 4: 388–389

27. Molloy SS, Thomas L, van Slyke JK, Stenberg PE, Thomas G (1994) Intracellular trafficking and activation of the furin proprotein convertase: localization to the TGN and recycling from the cell surface. EMBO J 13: 18–33

28. Murphy FA, Simpson DIH, Whitfield SG, Zlotnik I, Carter GB (1971) Marburg virus infection in monkeys. Lab Invest 24: 279–291

29. Ryabchikova EI, Kolesnikova LV, Tkachev VK, Pereboeva LA, Baranova SG, Rassadkin JN (1994) Ebola infection in four monkey species. Ninth International Conference on negative strand RNA viruses, Estoril, Portugal, p 164

30. Sanchez A, Kiley MP (1987) Identification and analysis of Ebola virus messenger RNA. Virology 157: 414–420

31. Sanchez A, Kiley MP, Holloway BP, Auperin DD (1993) Sequence analysis of the Ebola virus genome: organization, genetic elements, and comparison with the genome of Marburg virus. Virus Res 29: 215–240

32. Sanchez A, Yang ZY, Xu L, Nabel GJ, Crews T, Peters CJ (1998) Biochemical analysis of the secreted and virion glycoproteins of Ebola virus. J Virol 72: 6442–6447

33. Sanchez A, Trappier SG, Mahy BW, Peters CJ, Nichol ST (1996) The virion glycoprotein of Ebola viruses are encoded in two reading frames and are expressed through transcriptional editing. Proc Natl Acad Sci USA 93: 3602–3607

34. Schäfer W, Stroh A, Berghöfer S, Seiler J, Vey M, Kruse ML, Kern HF, Klenk H-D, Garten W (1995) Two independent targeting signals in the cytoplasmic domain determine trans-Golgi network localization and endosomal trafficking of the proprotein convertase furin. EMBO J 14: 2424–2435

35. Schnittler HJ, Mahner F, Drenckhahn D, Klenk H-D, Feldmann H (1993) Replication of Marburg virus in human endothelial cells. A possible mechanism for the development of viral hemorrhagic disease. J Clin Invest 91: 1301–1309

36. Schnittler H-J, Feldmann H (1999) Molecular pathogenesis of filovirus infections: role of macrophages and endothelial cells. Curr Topics Microbiol Immunol 235: 175–204

37. Seidah NG, Hamelin J, Mamarbachi M, Dong W, Tadro H, Mbikay M, Chretien M, Day R (1996) cDNA structure, tissue distribution, and chromosomal localization of rat PC7, a novel mammalian proprotein convertase closest to yeast kexin-like proteinases. Proc Natl Acad Sci USA 93: 3388–3393

38. Simpson DIH, Zlotnik I, Rutter DA (1968) Vervet monkey disease. Experimental infection of guinea pigs and monkeys with the causative agent. Br J Exp Pathol 49: 458–464

39. Skehel JJ, Bayley PM, Brown EB, Martin SR, Waterfield MD, White JM, Wilson IA, Wiley DC (1982) Changes in the conformation of influenza virus hemagglutinin at the pH optimum of virus-mediated membrane fusion. Proc Natl Acad Sci USA 79: 968–972

40. Takada A, Robison C, Goto H, Sanchez A, Murti KG, Whittl MA, Kawaoka Y (1997) A system for functional analysis of Ebola virus glycoprotein. Proc Natl Acad Sci USA 94: 14764–14769

41. Vey M, Schäfer W, Reis B, Ohuchi R, Britt W, Garten W, Klenk H-D, Radsak K (1995) Proteolytic processing of human cytomegalovirus glycoprotein B (gp UL55) is mediated by the human endoprotease furin. Virology 206: 746–749

42. Volchkov VE, Becker S, Volchkova VA, Ternovoj VA, Kotov AN, Netesov SV, Klenk H-D (1995) GP mRNA of Ebola virus is edited by the Ebola virus polymerase and by T7 and vaccinia virus polymerases. Virology 214: 421–430

43. Volchkov VE, Blinov VM, Netesov SV (1992) The envelope glycoprotein of Ebola virus contains an immunosuppressive like domain similar to oncogenic retovirus. FEBS Lett 305: 181–184

44. Volchkov VE, Blinov VM, Kotov AN, Chepurnov AA, Netesov SV (1993) The full-length nucleotide sequence of the Ebola virus. IXth International Congress of Virology, Glasgow, Scotland, P52–2
45. Volchkov VE, Feldmann H, Volchkova VA, Klenk H-D (1998a) Processing of the Ebola virus glycoprotein by the proprotein convertase furin. Proc Natl Acad Sci USA 95: 5 762–5 767
46. Volchkov VE, Volchkova VA, Slenczka W, Klenk H-D, Feldmann H (1998b) Release of viral glycoproteins during Ebola virus infection. Virology 245: 110–119
47. Weissenhorn W, Calder LJ, Wharton SA, Skehel JJ, Wiley D (1998) The central structural feature of the membrane fusion protein subunit from the Ebola virus glycoprotein is a long triple-stranded coiled coil. Proc Natl Acad Sci USA 95: 6 032–6 036
48. Weissenhorn W, Dessen A, Harrison SC, Skehel JJ, Wiley DC (1997) Atomic structure of the ectodomain from HIV-1 gp41. Nature 387: 426–430
49. Will C, Mühlberger E, Linder D, Slenczka W, Klenk H-D, Feldmann H (1993) Marburg virus gene four encodes the virion membrane protein, a type I transmembrane glycoprotein. J Virol 67: 1 203–1 210
50. Wise RJ, Barr PJ, Wong PA, Kiefer M, Brake AJ, Kaufman RJ (1990) Expression of a human proprotein processing enzyme: correct cleavage of the von Willebrand factor precursor at a paired basic amino acid site. Proc Natl Acad Sci USA 87: 9 378–9 382
51. Wool-Lewis RJ, Bates P (1998) Characterization of Ebola virus entry by using pseudo-typed viruses: identification of receptor-deficient cell lines. J Virol 72: 3 155–3 160
52. Yang Z, Delgado R, Xu L, Todd RF, Nabel EG, Sanchez A, Nabel GJ (1998) Distinct cellular interactions of secreted and transmembrane Ebola virus glycoproteins. Science 279: 1 034–1 036
53. Zaki SR, Peters CJ (1997) Viral hemorrhagic fevers. In: Connor DH, Chandler FW, Schwartz DA, Manz HJ, Lack EE (eds) The pathology of infectious diseases. Appleton and Lange, Norwalk, pp 347–364

Authors' address: Dr. H.-D. Klenk, Institut für Virologie, Philipps-Universität Marburg, Postfach 2360, D-35011 Marburg, Germany.

Retroviruses: ancient and modern

R. A. Weiss[1,2], **D. Griffiths**[1,2], **Y. Takeuchi**[1,2], **C. Patience**[1,3], and **P. J. W. Venables**[4]

[1]Institute of Cancer Research, London, U.K.
[2]Windeyer Institute of Medical Sciences, Royal Free & University College
London Medical School, London, U.K.
[3]BioTransplant Inc, Charlestown, Massachusetts, U.S.A.
[4]Kennedy Institute of Rheumatology, London, U.K.

Summary. Retroviruses are transmitted in two distinct ways: as infectious virions and as 'endogenous' proviral DNA integrated in the germ line of their hosts. Modern infectious viruses such as HIV recently infected mankind from simian hosts, whereas human endogenous retroviral genomes have been present throughout old world primate evolution. Recently we have characterised novel retroviruses in humans and pigs. Human retrovirus 5 (HRV-5) is detected as an exogenous genome in association with arthritis and systemic lupus erythematosus. Porcine endogenous retroviruses (PERV) are carried in swine DNA but can be activated to produce virions that are infectious for human cells, which has implications for xenotransplantation. A brief account of HRV-5 and PERV is given here.

Introduction

The *Retroviridae* [3] have been studied for much of the century since the pioneering work of Beijerinck on tobacco mosaic virus, and of Löffler and Frosch on foot-and-mouth disease distinguished viruses from other microbes. Swamp fever in horses was shown by Vallée and Carré in 1904 to be caused by a filterable agent, which we now know to be the lentivirus, equine infectious anaemia virus. Then retroviruses became focused on malignant disease: the discovery of avian leukosis virus in 1908 by Ellerman and Bang, avian sarcoma virus in 1911 by Rous, murine mammary tumour virus in 1936 by Bittner, murine leukaemia virus in 1951 by Gross, and human T-cell leukaemia virus in 1980 by Gallo. In this new era of AIDS, with the discovery of human immunodeficiency virus type 1 (HIV-1) in 1983 by Barré-Sinoussi, Chermann and Montagnier and HIV-2 in 1986 by Clavel and Montagnier, emphasis in retrovirus research has, of course, switched to immunopathology.

Aside from disease, retroviruses have provided important insights into and tools for molecular biology [3]. Without their study, for example, we would not have reverse transcriptase in order to make cDNA, we would have remained ignorant of the role of oncogenes in cancer for a much longer period, and we would not have such useful vectors for experimental and clinical gene transfer.

Retroviruses can be lethal pathogens, none more so than HIV-1. UNAIDS estimates that by December 1998, this virus had already infected some 50 million people worldwide and killed 14 million of them. HIV-1 is a thoroughly modern virus; it has probably arisen after a zoonotic infection from chimpanzees during the 20th century [5, 28] and raced through the human population, largely because of our sexual promiscuity. Other retroviruses cause ancient and asymptomatic infections, exemplified by that most intimate of host-parasite relations – endogenous retroviral genomes integrated into chromosomal DNA. These are vertically transmitted as inherited Mendelian traits. Human endogenous retroviruses (HERV) have been in humans for at least 50 million years [13, 18].

The contrasting modes of transmission of HIV and HERV highlight the different rates of evolution between an infectious, high turnover genome, and one embedded as integrated DNA in the germ line of the host. However, a common feature is that retroviruses can jump host species, sometimes across wide phylogenetic distances [14]. That is reason enough to treat the porcine endogenous retroviruses (PERV) described below with respect, as they can infect human cells when activated from pig chromosomal DNA as infectious virions.

Our laboratory has recently investigated two distinct types of retrovirus of potential but ill-understood impact for human health. The first is a group of endogenous C-type retroviruses in pigs [17], which are important in respect to the use of porcine cells and tissues for human xenotransplantation [27]. The second is a recently discovered but probably ancient virus called human retrovirus 5 (HRV-5) [9]. It is associated with arthritis and systemic lupus erythematosus [8], but the infection may be widespread.

Porcine endogenous retroviruses (PERV)

The release of PERV particles from lines of pig kidney cells has been known for over 25 years [2, 25]. With the growing interest in pig tissues and organs for transplantation into humans [27], we thought it was important to determine whether PERV particles might have a human host range.

Our initial study [17] showed that PERV released from the PK-15 line of pig kidney cells could infect certain human cells in culture, particularly the human embryonic kidney cell line 293. In contrast, PERV released from the mini-pig kidney cell line MPK appeared to infect only swine cells. Sequence analysis in *pol* showed that the PERVs of PK-15 and MPK were closely similar [17]. In collaboration with J. Stoye's group at the National Institute of Medical Research at Mill Hill, we detected two distinct envelope sequences, PERV-A and PERV-B, in 293 cells infected by PK-15 virions [12]. The virus released from MPK cells represents a third envelope type, PERV-C, which resembles an independently cloned genome [1]. The differences in host range of PERV-A, -B and -C can

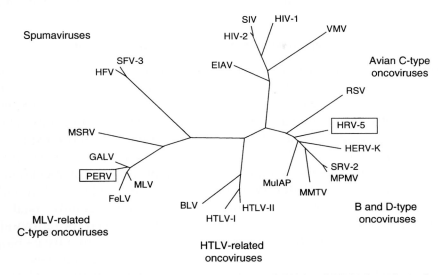

Fig. 1. Phylogenetic groups of retroviruses. PERV and HRV-5 are boxed

be attributed to their envelope glycoproteins that recognise distinct cell surface receptors [24].

The PERV genomes are most closely related to GALV and other mammalian C-type retroviruses (Fig. 1). PERV-A and -B are present as multiple copies in the normal DNA of all breeds of swine examined [12], though PERV-C may be more restricted. Thus, it would appear a difficult prospect to breed swine for use in human xenotransplantation which are free of PERV genomes. Provided the Mendelian genomes encoding potentially infectious PERV can be identified, their elimination might become feasible if genetic knockout technology for swine is developed.

The capacity of PERV strains to infect human cells in culture raises concern that these retroviruses might infect transplant recipients, and possibly be transmitted to these patients' contacts. The transplantation of swine tissue to an immunosuppressed human, and the generation of transgenic pigs bearing human genes to prevent hyperacute rejection of xenografts may heighten the risk of pig to human transmission [26]. These concerns have stimulated a retrospective analysis of patients who have been exposed to swine tissues. To date, published data are available on only 12 individuals – 10 diabetes patients transplanted with swine pancreatic islets [10], and 2 renal dialysis patients, whose circulation was linked for a short time period to swine kidneys extracorporeally [16]. In none of these 12 patients were PERV sequences found in peripheral blood samples taken months or years after their exposure to porcine tissue [10, 16]. It therefore appears that if PERVs are released from healthy porcine tissues, as seems likely from studies of short-term cultures of pig lymphocytes [29] and endothelial cells [15], they will not be highly infectious for humans. Nevertheless, close surveillance of patients and their contacts will be essential if xenotransplantation is to proceed. We would not wish to trigger a new retroviral pandemic via xenotransplantation.

Human retrovirus 5 and autoimmune disease

It has long been thought that retroviruses, as persistent infections, might play a role in the pathogenesis of autoimmune and chronic inflammatory diseases [7]. Particles resembling retroviruses have been seen in tissues from patients with Sjögren's syndrome [6, 30], rheumatoid arthritis [23] and psoriasis [11]. Recently, a human endogenous retroviral genome (HERV-K) closely related to those expressed in human testicular tumours [13] and the normal placenta [22], was also implicated in insulin dependent diabetes mellitus [4]. A different retroviral element related to the endogenous genomes HERV-W and HERV-9 has been identified in multiple sclerosis [19], but the significance of these findings remains controversial.

·Initially, we focused on Sjögren's syndrome and searched for retroviral particles by concentrating putative virions in sucrose gradients from tissue biopsies or short term cultures of affected salivary glands. We detected weak reverse transcriptase (RT) activity in fractions with the buoyant density typical (1.16 g/ml) of retrovirus particles. Using degenerate primers to retroviral *pol* (RT) and *pro* (protease) sequences, we amplified a sequence of 932 base pairs with open reading frames representing a novel retroviral sequence related to B-type and D-type retroviruses [9]. As HERV-K particles belong to the same subfamily of retroviruses (Fig. 1), we expected that this new retroviral element would be endogenous. However, when we examined normal human tissues and cell lines by Southern blotting and by PCR amplification using sequences specific to this genome, we found that it is not endogenous. We therefore assume that it must be transmitted via human-to-human infection, but we do not know as yet its prevalence and mode of transmission. We have provisionally named this novel retrovirus human retrovirus 5 (HRV-5), as it came to light after HTLV-1 and HTLV-2 and HIV-1 and HIV-2 [9].

As Fig. 1 indicates, HRV-5 is genetically most closely related to simian D-type retroviruses, such as Mason-Pfizer monkey virus, which cause immune deficiency, and to rodent intracisternal A-type particles and murine mammary tumour virus. We initially detected HRV-5 RNA in particles concentrated from tissues of patients with Sjögren's syndrome, normal salivary glands and lymphoma [9]. Further analysis based on PCR detection and amplification of DNA samples rather than RT-PCR of virion RNA have not supported a specific association of HRV-5 with Sjögren's syndrome or lymphoma [20, 21]. HRV-5 proviral DNA appears to be present at extremely low virus load; nested PCR is required for detection in most of the positive samples, giving a 2% frequency in the studies cited above. This may represent the prevalence of HRV-5 infection in the human samples studied, or it may be an underestimate due to the threshold of detection for a virus of such low load.

We have observed a significantly increased detection rate of HRV-5 in patients with arthritis and with systemic lupus erythematosus (SLE). Table 1 lists the

Table 1. Detection of human retrovirus 5 DNA in human tissue samples[a]

Tissue	Number tested	Number positive
Salivary gland		
Sjögren syndrome	86	1
Normal	9	0
Synovium		
Rheumatoid arthritis	25	12
Other arthritides	13	8
Normal	13	0
Lymph node		
Lymphoma	78	3
Non-malignant	64	1
Blood		
Rheumatoid arthritis	66	8
Systemic lupus erythematosus	69	11
Normal	103	1

[a]Data from [8, 20, 21]

frequency of detection in our recent survey [8]. Virus is often detected in the synovium of arthritic joints, but HRV-5 is not specific to rheumatoid arthritis, as it is also found in synovial tissue from patients with other arthropathies such as osteoarthritis.

We have recently been able to extend the HRV-5 genome sequence into the *gag* gene. By expressing Gag antigens it should be possible to develop serological assays for HRV-5 to be used for more detailed epidemiological and clinical studies.

In summary, we have evidence of human infection by a previously unknown retrovirus, HRV-5, which is distantly related to simian D-type retroviruses. The virus is present in affected tissues at very low frequency, probably less than one genome among 10,000 cells. So far, it is found most frequently in arthritides and SLE, but it would be unwise at this point to invoke a causative role. Nonetheless, the discovery of a new human retrovirus merits further study on its mode of transmission and possible pathogenesis. – These accounts of PERV and HRV-5 serve to remind us that novel retroviruses will continue to intrigue medical and veterinary virologists. Nevertheless, the most urgent problem is to control the HIV pandemic. With no safe, efficacious vaccine presently in sight, that is indeed a daunting task.

Acknowledgements

Our research of HIV, PERV and HRV-5 is supported by the Medical Research Council and the Arthritis Research Campaign.

References

1. Akiyoshi DE, Denaro M, Zhu H, Greenstein JL, Banerjee P, Fishman JA (1998) Identification of a full-length cDNA for an endogenous retrovirus of miniature swine. J Virol 72: 4503–4507

2. Armstrong JA, Porterfield JS, De Madrid AT (1971) C-type virus particles in pig kidney cell lines. J Gen Virol 10: 195–198

3. Coffin J, Hughes SH, Varmus HE (1997) Retroviruses. Cold Spring Harbor Laboratory Press, New York

4. Conrad B, Weissmahr RN, Boni J, Arcari R, Schüpbach J, Mach B (1997) A human endogenous retroviral superantigen as candidate autoimmune gene in type I diabetes. Cell 90: 303–313

5. Gao F, Bailes E, Robertson DL, Chen Y, Rodenburg CM, Michael SF, Cummins LB, Arthur LO, Peeters M, Shaw GM, Sharp PM, Hahn BH (1999) Origin of HIV-1 in the chimpanzee Pan troglodytes troglodytes. Nature 397: 436–441

6. Garry RF, Fermin CD, Hart DJ, Alexander SS, Donehower LA, Luo-Zhang H (1990) Detection of a human intracisternal A-type retroviral particle antigenically related to HIV. Science 250: 1127–1129

7. Garry RF, Kreig AM, Cheevers WP, Montelaro RC, Golding H, Fermin CD, Gallaher WR (1995) Retroviruses and their roles in chronic inflammatory diseases and autoimmunity. In: Levy JA (ed) The retroviridae, vol 4. Plenum Press, New York, pp 491–603

8. Griffiths DJ, Cooke SP, Herve C, Rigby SP, Mallon E, Hajeer A, Lock M, Emery V, Taylor P, Pantelidis P, Bunker CB, du Bois R, Weiss RA, Venables PJ (1999) Detection of human retrocirus 5 in patients with arthritis and systemic lupus erythematosus. Arthritis Rheum 42: 448–454

9. Griffiths DJ, Venables PJ, Weiss RA, Boyd MT (1997) A novel exogenuous retrovirus sequence identified in humans. J Virol 71: 2866–2872

10. Heneine W, Tibell A, Switzer WM, Sandstrom P, Rosales GV, Mathews A, Korsgren O, Chapman LE, Folks TM, Groth CG (1998) No evidence of infection with porcine endogenous retrovirus in recipients of porcine islet-cell xenografts. Lancet 352: 695–699

11. Iversen OJ (1990) The expression of retrovirus-like particles in psoriasis. J Invest Dermatol 95: 41S–43S

12. Le Tissier P, Stoye JP, Takeuchi Y, Patience C, Weiss RA (1997) Two sets of human-tropic pig retrovirus. Nature 389: 681–682

13. Löwer R, Löwer J, Kurth R (1996) The viruses in all of us: characteristics and biological significance of human endogenous retrovirus sequences. Proc Natl Acad Sci USA 93: 5177–5184

14. Martin J, Herniou E, Cook J, O'Neill RW, Tristem M (1999) Interclass transmission and phyletic host tracking in murine leukemia virus-related retroviruses. J Virol 73: 2442–2449

15. Martin U, Kiessig V, Blusch JH, Haverich A, von der Helm K, Herden T, Steinhoff G (1998) Expression of pig endogeneous retrovirus by primary porcine endothelial cells and infection of human cells. Lancet 352: 692–694

16. Patience C, Patton GS, Takeuchi Y, Weiss RA, McClure MO, Rydberg L, Breimer ME (1998) No evidence of pig DNA or retroviral infection in patients with short-term extracorporeal connection of pig kidneys. Lancet 352: 699–701

17. Patience C, Takeuchi Y, Weiss RA (1997) Infection of human cells by an endogenous retrovirus of pigs. Nature Med 3: 282–286

18. Patience, C, Wilkinson DA, Weiss RA (1997) Our retroviral heritage. Trends Genet 13: 116–120

19. Perron H, Garson JA, Bedin F, Beseme F, Paranhos-Baccala G, Komurian-Pradel F, Mallet F, Tuke PW, Voisset C, Blond JL, Lalande B, Seigneurin JM, Mandrand B, and the Collaborative Research Group on Multiple Sclerosis (1997) Molecular identification of a novel retrovirus repeatedly isolated from patients with multiple sclerosis. Proc Natl Acad Sci USA 94: 7 583–7 588

20. Rigby SP, Griffiths DJ, Jarrett RF, Weiss RA, Venables PJ (1998) A new human retrovirus: a role in lymphoma? Am J Med 104: 99–100

21. Rigby SP, Griffiths DJ, Weiss RA, Venables PJ (1997) Human retrovirus-5 proviral DNA is rarely detected in salivary gland biopsy tissues from patients with Sjögrens's syndrome. Arthritis Rheum 40: 2 016–2 021

22. Simpson GR, Patience C, Lower R, Tönjes RR, Moore HD, Weiss RA, Boyd MT (1996) Endogenous D-type (HERV-K) related sequences are packaged into retroviral particles in the placenta and possess open reading frames for reverse tanscriptase. Virology 222: 451–456

23. Stransky G, Vernon J, Aicher WK, Moreland LW, Gay RE, Gay S (1993) Virus-like particles in synovial fluids from patients with rheumatoid arthritis. Br J Rheumatol 32: 1 044–1 048

24. Takeuchi Y, Patience C, Magre S, Weiss RA, Banerjee PT, Le Tissier P, Stoye JP (1998) Host range and interference studies of three classes of pig endogenous retrovirus. J Virol 72: 9 986–9 991

25. Todaro GJ, Benveniste RE, Lieber MM, Sherr CJ (1974) Characterization of a type C virus released from the porcine cell line PK(15). Virology 58: 65–74

26. Weiss RA (1998) Transgenic pigs and virus adaptation. Nature 391: 327–328

27. Weiss RA (1998) Xenotransplantation. Br Med J 317: 931–934

28. Weiss RA, Wrangham RW (1999) The origin of HIV-1: From *Pan* to pandemic. Nature 397: 385–386

29. Wilson CA, Wong S, Muller J, Davidson CE, Rose TM, Burd P (1998) Type C retrovirus released from porcine primary peripheral blood mononuclear cells infects human cells. J Virol 72: 3 082–3 087

30. Yamano S, Renard JN, Mizuno F, Narita Y, Uchida Y, Higashiyama H, Sakurai H, Saito I (1997) Retrovirus in salivary glands from patients with Sjogren's syndrome. J Clin Pathol 50: 223–230

Authors' address: Prof. R. A. Weiss, Windeyer Institute of Medical Sciences, University College London, 46 Cleveland Street, London W1P 6DB, U.K.

Foot-and-mouth disease and beyond: vaccine design, past, present and future

F. Brown

Plum Animal Disease Center, Greenport, New York, U.S.A.

Summary. The first experimental vaccines against foot-and-mouth disease were made in 1925 by Vallee, Carre and Rinjard using formaldehyde inactivation of tongue tissue from cattle infected with the virus. This method was essentially unaltered until the late 1940s when the important experiments by Frenkel in Holland showed that the quantities of virus required for vaccine production could be obtained from fragments of tongue epithelium incubated in vitro following infection with the virus. This major step made possible the comprehensive vaccination programmes which followed in Western Europe and which, in turn, resulted in the elimination of the disease from that part of the world by 1989. This spectacular success has led many to question whether other kinds of vaccine are required to control the disease worldwide. Such reservations ignore the danger to the environment associated with the growth of large amounts of virus. This can never be a zero-risk situation. Consequently, a vaccine which is not based on infectious virus as starting material has many attractions from safety considerations alone. In addition, a vaccine based on more fundamental considerations would not only be more aesthetically satisfying but could possibly provide an understanding at the molecular level of antigenic variation, still a problem in the control of the disease.

The advances in our knowledge of the structure of the virus and the fragments which elicit a protective immune response now allow us to envisage a vaccine which does not require infectious virus and which protects against the multiple serotypes of the agent. Since antigenic variation is still a major problem in the control of the disease by vaccination, such a product would have important advantages over the current vaccines.

Introduction

Celebrating the momentous observation by Loeffler and Frosch [25] that an infectious animal disease could be caused by a filterable agent, one should appropriately start on an historic note, by referring to a meeting on foot-and-mouth disease which also took place in Germany. At a meeting of the Berlin Microbiological

Society on 27 April 1931, there were four presentations on foot-and-mouth disease, all by scientists from Insel Riems. The topics in order of presentation were:

i.   K. Trautwein, The plurality of the FMD virus;
ii. F. Hecke, Cultivating the FMD virus;
iii. G. Pyl, Concerning methods on concentration of the FMD virus from virus-containing substrates;
iv. O. Waldmann, The experimental proof of permanent carriers in FMD.

In the intervening 67 years, topic (ii), which provided the basic information for vaccine production, has been solved. We now know that the virus can be grown extremely well outside the animal body. Following the pioneering work of Hecke [17], and the Maitlands [26] in England, the major step was made by Frenkel [16] in Holland. By growing the virus in surviving bovine tongue epithelial cells, and helped in no small measure by the antibiotics that had recently become available, Frenkel was able to produce the quantities needed for vaccination to be undertaken on a major scale. Although it had been shown by Rosenbusch et al. [33] that large-scale production of vaccines could be achieved by growing the virus in animals, it was clear that vaccination against the disease on a worldwide scale would require the use of viruses produced in vitro.

Following the lead provided by the Dutch in 1952, Germany, France and other countries in Western Europe introduced comprehensive vaccination programmes which were to result in control of the disease in that area. Clearly some problems were encountered but once it had been accepted that inactivation of the virus with formaldehyde was incomplete and its use had been superseded by imines [3, 11], these hurdles disappeared. Moosbrugger [28] was the first to suspect, in 1948, that inactivation with formaldehyde was incomplete and independent studies by Brown et al. [10] and Graves [20] provided experimental confirmation of this view. But it required the evidence from the analysis by molecular techniques [4, 22] of the RNA from viruses causing outbreaks in Europe to finally convince the diehards of the inefficiency of formaldehyde as an inactivant. Ironically, the overwhelming success of the vaccination programmes in Western Europe persuaded the European Economic Community to discontinue vaccination against FMD from 1992. Whether this was a wise decision remains to be seen since the disease is still present in many countries, some of which adjoin Western Europe.

The presentation of topic (iii) by Pyl at the Berlin meeting in 1931 was the forerunner of many attempts to purify and characterise the virus. Pure virus was first obtained by Brown and Cartwright in 1963 [9], thus allowing it to be analysed and characterised by modern molecular methods. Now we have crystals of the virus, its structure is known at 2.9 Å resolution [1], and its lifestyle is being studied in the most intimate detail. Who would have guessed that the particle contains actin [21, 29], proteins of the replication complex [29], and an enzyme, which can hydrolyse the viral RNA if the conditions are right (or wrong!) and thus allow it to commit suicide [13, 30].

The structural studies have provided us with information which explains why the virus is unstable below pH7 [12, 39] and why its density in caesium chloride is

Table 1. Antigenic components of foot and mouth disease virus

Component	Sedimentation constant	Composition
Virus particle	146S	1 molecule ssRNA ($M_r = 2.6 \times 10^6$), 60 copies of each of VP1-3 ($M_r c.\ 24 \times 10^3$) and VP4 ($M_r c.\ 8 \times 10^3$)
Empty particle	75S	60 copies of VP0[a], VP1 and VP3
Protein subunit	12S	Pentamer of VP1-3
Virus infection-associated antigen	3.8S	RNA polymerase ($M_r c.\ 56 \times 10^3$)

[a] VP0 comprises VP4 and VP2 covalently linked

so much higher, at 1.44g/cc, than that of the structurally similar poliovirus, which has the considerably lower value of 1.34g/cc. It has also given us the opportunity to study the immunogenic structure of the virus particle and the chemical basis for its antigenic variation.

Although it was appreciated that harvests of virus grown in tissue culture cells and suitably inactivated gave good protective immunity, it was clearly desirable to determine which components in the harvests were responsible for providing this immunity. Analysis of the harvests showed that, in addition to the infectious virus particles, three additional virus-specific components are present (Table 1). The virus particle, which is the major immunogenic component, consists of a copy of single-stranded, positive-sense RNA encapsidated into a 30 nm icosahedral particle with 60 copies of each of four proteins (VP1, VP2 and VP3, mol. wt. c. 24×10^3 and VP4 c.8×10^3). The 75S component, which is similar in size and shape, does not contain any RNA but it does contain the same four proteins, with VP4 and VP0 covalently linked. This particle is also immunogenic provided its fragile structure is preserved [35]. In contrast, the 12S component, a pentamer of VP1, VP2, and VP3, is poorly immunogenic. The virus infection-associated antigen, which is now recognised to be the virus RNA polymerase, has no capacity to elicit neutralising antibodies.

Evidence that VP1 has a dominant role in eliciting neutralising antibodies was provided by Brown et al. in 1967 [43] when they found that particles of a virus of serotype O which had been treated with trypsin elicited only low levels of neutralising antibodies. This loss of activity was caused by cleavage of VP1. The other capsid proteins were unaffected (Fig. 1). Moreover, antibodies of the IgM class no longer complexed with the treated particles whereas they attached at specific sites on the untreated virus. The importance of VP1 was confirmed by Laporte et al. in 1973 [24] when they showed that the isolated protein elicited neutralising antibodies in pigs. This immunogenic activity was also described by Bachrach et al. in 1975 [2] but the response was very much lower than that obtained with inactivated virus particles. Similarly, the protein obtained by expressing the gene coding for it was also poorly immunogenic [23]. These observations indicate that the configuration of the isolated VP1 is

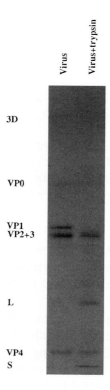

Fig. 1. PAGE analysis of ^{35}S-methionine labelled proteins from virus particles and trypsin treated virus particles

different from that which it possesses when it forms part of the virus capsid structure.

An alternative approach was to identify the immunising epitopes on VP1. This information was obtained by four groups in the 1980s, using different methods [6, 19, 31, 37]. The first studies to identify immunogenic fragments, made by Strohmaier et al. [37], followed the classical approach described by Anderer with tobacco mosaic virus, namely fragmentation of VP1 by cyanogen bromide or proteolytic enzymes, followed by testing of the individual cleavage products for immunogenic activity. Although the activity of the fragments was extremely low, Strohmaier and his colleagues identified two potential immunogenic sites, at residues 146–154 and 201–213.

In a different approach, Bittle and his colleagues [6] took advantage of the antigenic variability of the virus and reasoned that, since this variability would be reflected in amino acid sequence variation, a comparison of the sequences of viruses belonging to different serotypes would pinpoint potentially important sites. By comparing the derived amino acid sequences of VP1 from four isolates belonging to three serotypes (which were the only sequences available at that time), they identified three regions of considerable variability, at positions within 41–60, 138–160 and 194–205. Direct testing of the immunogenic activity of 20-mer peptides (1–20, 21–40 etc.) encompassing the entire sequence of VP1 from a virus of serotype O demonstrated that one inoculation of the 141–160 sequence, linked via an added cysteine at the C-terminus to keyhole limpet haemocyanin, elicited levels of neutralising antibody which afforded protection of guinea pigs against challenge infection with $10^4 ID_{50}$ of the homologous virus. The 201–213 sequence at the C-terminus also elicited neutralising antibody but the levels were very much lower than those obtained with the 141–160 sequence. No other regions elicited measurable levels of neutralising antibody. A predictive approach by Pfaff et al. [31], based on the reasoning that a good candidate structure would be strongly helical with hydrophilic and hydrophobic zones on opposite sides of the helix, concluded that residues 144–159 would meet this requirement.

The 'pepscan' method introduced by Geysen et al. [19], which has proved useful in identifying continuous epitopes on several protein molecules, has also been used to identify immunogenic sites on FMD virus. This method is based on the reaction of overlapping hexapeptides comprising the entire sequence of a protein with antibody elicited by the protein. In the case of FMD virus, the peptides comprising VP1 were screened with antibody from hyperimmunised animals. Those peptides that reacted were detected by a second screening step with an antispecies antiserum directed against the hyperimmune serum. This approach also pinpointed the immunodominance of a region within the 141–160 sequence. The fact that all four approaches pointed to the immunodominance of this region of VP1 has resulted in a concentration of effort on this sequence. Moreover, subsequent examination of the virus by X-ray crystallography [1] has revealed that the immunogenic sequence forms part of a highly disordered loop, which comprises residues 134–158. Since it is generally accepted that flexible regions of a protein are more immunogenic, the structural studies lend strong support to the biological observations. Moreover, the C-terminus of VP1 is located close to the loop region, providing structural evidence for the enhanced immunogenicity of the hybrid peptide comprising residues 141–160 and 200–213 [14].

Experiments with many more virus isolates have confirmed that the 138–160 sequence is highly variable in all seven serotypes. Not only is the sequence variable but it also varies in length. Evidence has now been provided which shows that the region in isolates from all serotypes will elicit high levels of neutralising antibody.

The early results in guinea pigs provided hope that a peptide vaccine for FMD would soon become available. Although there is still no product, there are several reasons why the peptide approach should be pursued. For example, Taboga et al. [38], in a large-scale trial in cattle, were able to protect about 40% of the animals. Of the unprotected animals, the disease in 12 out of 29 was caused by a variant in the challenge virus.

Chemical synthesis allows the production of stable products, which are not bedevilled by the problems associated with materials produced by biological procedures. Moreover, the ease with which they can be synthesised has made this approach feasible. Peptide vaccines also offer the advantage that their simple structure, compared with that of proteins, allows them to be manipulated more readily. Probably of more importance, however, is the fact that as we learn more about the immune response at the molecular level, it becomes increasingly obvious that the interaction of the MHC molecules with proteins involves only short amino acid sequences.

One of the more interesting innovations over the past few years has been the demonstration that peptides prepared from D-amino acids (retro-inverso peptides) elicit levels of antibodies in guinea pigs that are higher than those obtained with the corresponding L-peptides [7]. Moreover, the response is longer lasting. This approach has considerable potential if the results in guinea pigs can be reproduced in cattle and pigs.

The importance of T-cell epitopes

If we are to realise the goal of a synthetic vaccine, it is necessary to identify a T-cell epitope suitable for the host species, which can be linked to the specific B-cell epitope. This has been achieved for experimental models, namely mice of the H-2d haplotype [15] and guinea pigs (unpubl. obs.). Moreover, Hensen's group has identified a T-cell epitope on VP4 of the virus particle, which is suitable for cattle [41]. These observations indicate that a synthetic peptide vaccine is within reach.

The importance of antigenic variation in FMD has been recognised since the 1920s [40, 42]. This plurality, as Trautwein described it in 1931, is of great importance in the control of disease by vaccination and, even in 1998, it still represents a considerable threat. We now recognise seven serotypes – remarkably no more have been found since 1954. Vaccines for each serotype have performed well under field conditions but there is the insidious occurrence of sub-types within the serotypes. This can, and frequently does, cause problems in the control of the disease by vaccination.

The occurrence of antigenic variants within a serotype has also been known since the 1920s [5]. This situation was described by Trautwein at the Berlin meeting in 1931 and was demonstrated dramatically in the major Mexican outbreak during 1946–1954 [18]. In closer terms geographically, antigenic variation in the virus of serotype O led to the need to change this valency of the vaccines being used so successfully in Europe during the early 1960s. The appearance of a virus of sub-type 1 and the names O-Kaufbeuren, O-Lausanne and O-BFS are well known to those working with the virus at that time.

A molecular understanding of antigenic variation started to emerge in the early 1980s, in the years immediately following the sequencing of the viral RNA of different serotypes of the virus. The importance of VP1, and in particular the GH loop, in eliciting the production of neutralising antibodies, has been described above. As mentioned earlier, comparison of the amino acid sequences of this loop region from many isolates of the virus has shown that these differ considerably between serotypes and to a lesser extent between individual isolates of a single serotype. This type of analysis has also demonstrated very clearly that field isolates often consist of a mixture of antigenic variants, which differ at one or more amino acid positions on the loop. The first demonstration of this situation was found in experiments with the A12 virus which had been used in extensive vaccination experiments at the Pirbright laboratory in England. This virus, which had been isolated from an outbreak of the disease in England in 1932, was given to the Plum Island Animal Disease Centre in the U.S.A. when experimental studies on the disease were started there in 1954. In extensive studies with this virus, several antigenic variants were isolated [8, 34]. One of these variants was sequenced when this methodology became available in the late 1970s and, on the basis of this information, a 20-mer peptide corresponding to the amino acids at positions 141–160 of VP1 (the same positions in the serotype O studies referred to above) was used to immunise guinea pigs. The antibody neutralised the virus whose sequence had been determined, but did not neutralise the parent virus. It was eventually recognised that this was because the parent virus was a mixture [34]. Subsequent analysis of the viruses which had been "derived" at Plum Island showed that amino acid substitutions at positions 148 and 153 on VP1 were sufficient to account for considerable variation because the sequence of the remainder of the capsid proteins was the same in each variant [8].

The structural studies showed that the 20-mer peptide was located on a prominent loop region of the virus, which was so mobile that its structure could not be determined [1]. This loop is probably incidental to the architecture of the virus particle because it can be removed without altering its sedimentation constant or appearance in the electron microscope. As a consequence, it seems probable that the chance of survival of antigenic variants is enhanced because the variation will not affect the assembly of viable particles. In contrast, we could cite the lack of variation in poliovirus where the viruses used for the production of vaccines in the 1950s are still used in the 1990s.

The situation with a virus of serotype 0–1 is not so straightforward. For example, antigenic variants obtained by growing the virus in the presence of a monoclonal antibody against the GH loop region were found to have the same sequence in this loop as the parent virus. Interestingly, the changes were found at residues 43, 48 and 59 of the BC loop of VP1 [32]. X-ray crystallographic studies of two of the mutants, (i) Thr 43-> Ala and (ii) His 59-> Tyr, showed that the substitutions had caused a switch in the configuration of the GH loop. This loop has two configurations on the virus particle. In the predominant configuration the GH loop overlays the BC loop in the region of residues 43 to 59. This is the conformation recognised by the antibody. The mutants escape neutralisation through the substituted residues in the BC loop interacting with the GH loop to destabilise its predominant conformation, thus destroying the integrity of the epitope recognised by the antibody. This mechanism is similar to the allosteric switching of enzymes.

The fourth topic at the Berlin meeting was that presented by Waldmann on carrier animals. Waldmann provided evidence that virus could be recovered from the blood and urine of cattle as long as 158 and 246 days respectivley after infection. Subsequent work by several groups has amply confirmed the existence of the carrier state in cattle (see review by Salt [36]). But there is still much debate on whether this is important in the transmission of the disease despite the results described by Waldmann in 1931. There is, however, good evidence that the virus can be transmitted from buffalo to cattle [36].

With the exception of one short report that the wild boar can be a carrier, it is generally cosidered that the virus does not persist in pigs. However, Mezencio et al.

[27] have recently obtained evidence for the persistence of the virus in pigs. During experiments designed to distinguish between convalescent animals and those that had been immunised with the classical inactivated vaccine, it was observed that the level of antibodies against the non-structural proteins 2C, 3 ABC and 3D, and the structural protein precursor P1 in the sera of both convalescent cattle and pigs fluctuated significantly. Judging that this fluctuation was probably caused by repeated stimulation of the immune system by virus replication, the RNA was extracted from sera taken at frequent intervals after infection, and amplified using primers corresponding to the non-structural protein 2B. A specific band was obtained with those samples obtained from sera collected a few days before the rise in antibody titres. This temporal relationship between the increase in viral RNA and the rise in specific antibodies to the non-structural proteins provides good evidence for the persistence of the virus in pigs.

Conclusion

Summing up, one should ask whether all the information we have accumulated in the past 100 years, since Loeffler and Frosch showed that the enemy is a virus, allows us to consider the eradication of the disease seriously.

1. Effective vaccines are available and have been used with great success, as demonstrated by the control of the disease in Western Europe and some countries in South America.

2. The overt disease can be diagnosed very rapidly.

3. Animals which have become infected but no longer show signs of the disease, although still carrying the virus, can be identified by simple serological tests. Hence, their movement could be controlled so that they could not become the source of further outbreaks.

It would be fitting on this 100th Anniversary if a programme of world-wide eradication could be initiated because, while the disease exists anywhere in the world, any country is at risk.

References

1. Acharya R, Fry E, Stuart D, Fox G, Rowlands DJ, Brown F (1989) The three-dimensional structure of foot-and-mouth disease virus at 2.9 Å resolution. Nature 337: 709–716
2. Bachrach HL, Moore DM, McKercher PD, Polatnick J (1975) Immune and antibody responses to an isolated capsid protein of foot-and-mouth disease virus. J Immunol 115: 1636–1641
3. Bahnemann HG (1975) Binary ethyleneimine as an inactivant for foot-and-mouth disease virus and its application for vaccine production. Arch Virol 47: 47–56
4. Beck E, Strohmaier K (1987) Subtyping of European foot-and-mouth disease virus strains by nucleotide sequence determination. J Virol 61: 1621–1629
5. Bedson SP, Maitland HB, Burbury YM (1927) Foot-and-Mouth Disease Research Committee, 2nd Report. HMSO, London, p 99
6. Bittle JL, Houghten RA, Alexander H, Shinnick TM, Sutcliffe JG, Lerner RA, Rowlands DJ, Brown F (1982) Protection against foot-and-mouth disease by immunization with

a chemically synthesized peptide predicted from the viral nucleotide sequence. Nature 298: 30–33

7. Braind J-P, Benkirane N, Guichard G, Newman JFE, Van Regenmortel MHV, Brown F, Muller S (1997) A retro-inverso peptide corresponding to the GH loop of foot-and-mouth disease virus elicits high levels of long-lasting protective neutralising antibodies. Proc Natl Acad Sci USA 94: 12 445–12 550

8. Brown F (1994) The importance of antigenic variation in vaccine design. Arch Virol [Suppl] 9: 1–8

9. Brown F, Cartwright B (1963) Purification of radio-active foot-and-mouth disease virus. Nature 31: 1168–1170

10. Brown F, Hyslop NSt.G, Crick J, Morrow AW (1963) The use of acetylethyleneimine in the production of inactivated foot-and-mouth vaccines. J Hyg 61: 337–344

11. Brown F, Crick J (1959) Application of agar gel diffusion analysis to a study of the antigenic structure of inactivated vaccines prepared from the virus of foot-and-mouth disease. J Immunol 82: 444–447

12. Curry S, Abrams CC, Fry E, Crowther JC, Belsham GJ Stuart DI, King AM (1995) Viral RNA modulates the acid sensitivity of foot-and-mouth disease virus capsids. J Virol 69: 430–438

13. Denoya CD, Scodeller EA, Vasquez C, La Torre JL (1978) Foot-and-mouth disease virus: II. Endoribonuclease activity within purified virions. Virology 89: 67–74

14. DiMarchi R, Brooke G, Gale C, Cracknell V, Doel T, Mowat N (1986) Protection of cattle against foot-and-mouth disease by a synthetic peptide. Science 232: 639–641

15. Francis MJ, Hastings GZ, Syred AD, McGinn B, Brown F, Rowlands DJ (1987) Non-responsiveness to a foot-and-mouth disease virus peptide overcome by addition of foreign helper T-cell determinants. Nature 330: 168–170

16. Frenkel HS (1930) La culture du virus de la fievre aphteuse sur l'epithelium de la langue des bovides. Bull Off Int Epizool 28: 155–162

17. Hecke F (1931) Zuechtungsversuche des Maul-und-Klauenseuchevirus in Gewebekulturen. Zbl Bakt Parasitk I 116: 386–414

18. Galloway IA, Henderson WM, Brooksby JB (1948) Strains of the virus of foot-and-mouth disease recovered form outbreaks in Mexico. Proc Soc Exp Biol Med 69: 57–63

19. Geysen HM, Meloen RH, Barteling SJ (1984) Use of peptide synthesis to probe viral antigens for epitopes to a resolution of a single amino acid. Proc Natl Acad Sci USA 81: 3998–4002

20. Graves JH (1963) Formaldehyde inactivation of foot-and-mouth disease virus as applied to vaccine preparation. Am J Vet Res 24: 1 131–1 135

21. Grigera PR, Tisminetzky SG, Lebendiker MB, Periolo OH, La Torre JL (1988) Presence of a 43-KDa host-cell polypeptide in purified aphthovirions. Virology 165: 584–588

22. King AMQ, Underwood BO, McCahon D, Newman JWI, Brown F (1981) Biochemical identification of viruses causing the 1981 outbreaks of foot-and-mouth disease in the U.K. Nature 293: 479–480

23. Kleid DG, Yansura D, Small B, Dowbenko D, Moore DM, Grubman MJ, McKercher PD, Morgan DO, Robertson BH, Bachrach HL (1981) Cloned viral protein vaccine for foot-and-mouth disease; responses in cattle and swine. Science 214: 1125–1129

24. Laporte J, Grosclaude J, Wantyghem J, Bernard S, Rouze P (1973) Neutralisation en culture cellulaire du pouvoir infectieux du virus de la fievre aphteuse par des serum provenant de porcs immunises a l'aide d'une proteine virale purifiee. C R Acad Sci 276: 3399–3401

25. Loeffler F, Frosch P (1898) Berichte der Kommission zur Erforschung der Maul-und-

Klauenseuche bei dem Institut fur Infektionskrankbeiten in Berlin. Centralblatt Bakt I Abt Orig 23: 371–391

26. Maitland MC, Maitland HB (1931) Cultivation of foot-and-mouth disease virus. J Comp Pathol Ther 44: 106–113

27. Mezencio JMS, Babcock GD, Kramer E, Brown F (1999) Evidence for the persistence of foot-and-mouth disease virus in pigs. Vet J 157: 213–217

28. Moosbrugger GA (1948) Rechserches experimentales sur la fievre aphteuse. Schweiz Arch Tierheilk 90: 179–198

29. Newman JFE, Brown F (1997) Foot-and-mouth disease virus and poliovirus particles contain proteins of the replication complex. J Virol 71: 7657–7662

30. Newman, JFE, Piatti PG, Gorman BM, Burrage TG, Ryan MD, Flint M, Brown F (1994) Foot-and-mouth disease virus particles contain replicase protein 3D. Proc Natl Acad Sci USA 91: 733–737

31. Pfaff E, Mussgay M, Bohm HO, Schulz GE, Schaller H (1982) Antibodies against a preselected peptide recognizes and neutralize foot-and-mouth disease virus. Eur Mol Biol Organ 1: 869–874

32. Parry N, Fox G, Rowlands D, Brown F, Fry E, Acharya R, Logan D, Stuart D (1990) Serological and structural evidence for a novel mechanism of antigenic variation in foot-and-mouth disease virus. Nature 347: 569–572

33. Rosenbusch CT, Decamps A, Gelormini N (1948) Intradermal foot-and-mouth disease vaccine. Results obtained from the first million head of cattle vaccinated. J Am Vet Med Assoc 112: 45–47

34. Rowlands DJ, Clarke BE, Carroll AR, Brown F, Nicholson BH, Bittle JL, Houghten RA, Lerner RA (1983) Chemical basis of antigenic variation in foot-and-mouth disease virus. Nature 306: 694–697

35. Rowlands DJ, Sangar DV, Brown F (1972) Stabilizing the immunizing antigen of foot-and-mouth disease virus by fixing with formaldehyde. Arch Ges Virusforsch 39: 274–283

36. Salt JS (1993) The carrier state in foot-and-mouth disease – an immunological review. Br Vet J 149: 207–223

37. Strohmaier K, Franze R, Adam K-H (1982) Localisation and characterisation of the antigenic portion of the foot-and-mouth disease virus protein. J Gen Virol 59: 295–306

38. Taboga O, Tami C, Carrillo E, Nunez JI, Rodriguez A, Saiz JC, Blanco E, Valero M-L, Roig A, Camarero JA, Andreu D, Mateu MG, Giralt E, Domingo E, Sobrino F, Palma EL (1997) A large-scale evaluation of peptide vaccines against foot-and-mouth disease: Lack of solid protection in cattle and isolation of escape mutants. J Virol 71: 2606–2614

39. Twomey T, France LL, Hassard S, Burrage TG, Newman JFE, Brown F (1995) Characterization of an acid-resistant mutant of foot-and-mouth disease virus. Virology 206: 69–75

40. Vallee H, Carre H (1922) Sur la pluralite du virus aphteux. C R Acad Sci 174: 1498–1500

41. van Lierop MJ, Wagenaar JP, van Noort JM, Hensen EJ (1995) Sequences derived from the highly antigenic VP1 region 140 to 160 of foot-and-mouth disease virus do not prime for a bovine T-cell response against intact virus. J Virol 69: 4511–4514

42. Waldmann O, Trautwein K (1926) Experimentelle Untersuchungen über die Pluralität des Maul-und Klauenseuche Virus. Berl Tierarztl Wochenschr 42: 569–571

43. Wild TF, Burroughs JN, Brown F (1969) Surface structure of foot-and-mouth disease virus. J Gen Virol 4: 313–320

Authors' address: Dr. F. Brown, Plum Animal Disease Center, P.O. Box 848, Greenport, 11944 NY, U.S.A.

Viruses and gene silencing in plants

D. Baulcombe

The Sainsbury Laboratory, John Innes Centre, Norwich, U.K.

Summary. Genetic engineering of virus resistance in plants may be conferred by transgenes based on sequences from the viral genome. In many instances the underlying mechanism involves the transgenically expressed proteins. However there are other examples in which the mechanism is based on RNA. It appears that this mechanism is related to post transcriptional gene silencing in transgenic plants. This gene silencing is likely to involve antisense RNA produced by the action of a host-encoded RNA dependent RNA polymerase. The natural role of this mechanism is as a genetic immune system conferring protection against viruses. There may also be a genomic role of the process reflected in RNA directed methylation of transgenes. Further understanding of this mechanism has obvious implications for virus resistance in plants. In addition the gene silencing can be used as a component of a new technology with application in functional genomics.

Introduction

Viruses are important in their own right as agents of disease in plants and animals. However they are also important as probes of biological systems. Many fundamental processes have been revealed using viruses and virus-based enabling technologies have contributed to discoveries in biology. In this paper I describe an emerging story in plant virology that fits this pattern in several respects.

Genetic engineering of virus resistance

The first approach to genetic engineering of virus resistance in plants involved transgenic expression of the tobacco mosaic virus (TMV) coat protein [1]. This first approach was remarkably successful although, more than ten years later, the precise mechanism is not fully understood. One possibility is that the transgenic coat protein inhibits disassembly of TMV particles in the initially infected cell [36]. However, from the available information, it is also possible that the transgenic protein blocks virion receptor sites or is an elicitor of host defense [15].

Spurred on by the initial success with TMV there were numerous attempts to generate coat protein mediated protection against other viruses [4, 9, 20, 48]. There have also attempts to genetically engineer virus resistance by transgenic expression of viral genes other than the coat protein gene [3, 7, 17, 32]. It was reasoned, for example, that if transgenic expression of the coat protein could confer resistance by affecting the virions then expression of the replication enzyme might affect virus replication. In the same way, expression of the movement protein might affect viral movement.

Many of these exercises were successful in that they produced virus-resistant transgenic plants. However the detailed analysis in some of the lines revealed two significant features that were not easily explained by a protein-based mechanism [5]. First there could be resistance with transgenes specifying a non translatable RNA. The second anomaly was the finding that the level of resistance did not correlate with the expression of the transgene. In many instances the low level transgene expressers were more resistant than high level expressers of the same transgene. Neither of these anomalous findings was consistent with protein-based models of the resistance mechanism. It seemed that the mechanism was fundamentally different from that operating in the coat protein transgenic plants [5].

The first clue to the explanation of these anomalous examples was from plants carrying a tobacco etch virus (TEV) transgene [31]. The resistance in some of these lines was induced following TEV infection and, associated with the onset of resistance, the levels of the TEV transgene RNA declined. Based on these results it was suggested that a single mechanism could account for both the resistance and the suppression of the transgene RNA. If that were the case, because TEV is a cytoplasmically replicating virus with an RNA genome, the mechanism would operate in the cytoplasm at the level of RNA [31].

This hypothetical cytoplasmic mechanism (Fig. 1) could also account for the inverse relationship between transgene expression level and virus resistance in plants with transgenes based on PVX [32], potato virus Y [30] and other viruses. To test this possibility a series of crosses were made with lines carrying transgenes encoding the replication enzyme of potato virus X (PVX) [34]. Some of the parents in these crosses were virus-resistant lines expressing the PVX transgene at a low level. The other parents were virus susceptible virus lines carrying exactly the same PVX transgene construct (Fig. 2). The PVX transgene RNA was abundant in these lines. In the progeny carrying transgenes from both parents there were only low levels of the viral RNA and the plants were resistant against PVX. These data were completely consistent with a single mechanism causing low accumulation of transgene RNA and virus resistance [34]. Subsequently it transpires that many, although not all examples of virus resistance due to virus-derived transgenes are due to this RNA-mediated mechanism [5].

Post transcriptional gene silencing

These results were not the only anomalous findings involving suppression of transgene expression. For example, expression of a non-viral transgene could be suppressed by re-transformation with a second homologous transgene [33]. Also, when plants were transformed with homologues of endogenous genes the transgene and the endogenous gene were co-suppressed in some of the lines [25, 35].

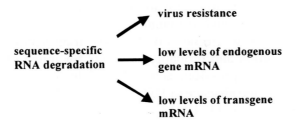

Fig. 1. The relationship of gene silencing and virus resistance. It is proposed that an RNA degradation mechanism is given specificity by antisense RNA (see text). This RNA degradation mechanism is able to target viral RNA to produce resistance against viruses that are similar to the transgene at the nucleotide sequence level. The same mechanism is also able to degrade the RNA product of the transgene so that there is only a low level expression of the transgene. This mechanism can also be targeted against the RNA of endogenous genes so that the endogenous gene and the transgene are cosuppressed

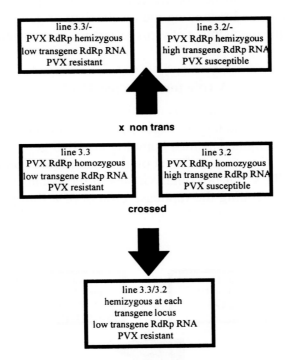

Fig. 2. Experimental testing of the relationship of virus resistance and PTGS. Lines 3.3 and 3.2 are tobacco lines carrying a PVX RdRp transgene. The diagram illustrates the phenotypes of plants carrying these genes either individually or in combination and shows how it was concluded that the PVX RdRp transgenes conferring PVX resistance were also able to confer PTGS. The primary data are presented previously [34]. The crosses with non-transformed plants (upper panel) revealed that the transgene phenotype with these lines was not affected by transgene dosage. The crosses between the two lines showed that the transgene in line 3.3 could suppress expression of the transgene in line 3.2. The results also showed that PVX resistance and low level transgene expression were epistatic to high level expression and PVX susceptibility

These examples of gene silencing are now known to be either transcriptional or post-transcriptional although it remains an open question as to whether there are mechanistic links between the two. The post-transcriptional gene silencing (PTGS) is more common and probably accounts for much of the between-line variation in transgene expression: active PTGS leads to low level transgene expression whereas, in the absence of PTGS or if the PTGS is weak, the transgene is expressed at a high level [11, 12, 14, 21, 22].

It seemed likely that the PTGS of non-viral transgenes could be related to the cytoplasmic mechanism of RNA turnover that had been invoked as an explanation of the transgenic virus resistance. To test this idea we inoculated PVX vectors to tobacco lines exhibiting PTGS of GUS [14]. These PVX constructs carrying non-GUS inserts were able to replicate to a high level. In contrast, the PVX vectors with GUS inserts were not able to accumulate or spread on the GUS silencing lines. In effect the GUS transgene in these lines was a virus resistance gene (Fig. 3) and the link was established between PTGS and transgenic virus resistance [14]. It was clear that understanding the mechanism of gene silencing would be informative about the mechanism of virus resistance in transgenic plants and vice versa.

A role for antisense RNA

Post transcriptional gene silencing is highly nucleotide sequence specific: silencing of chalcone synthase is targeted against other chalcone synthase genes [35]; silencing of GUS is targeted against homologous GUS genes [14, 21, 22] and so on. Similarly a PTGS transgene based on the coat protein gene of TEV confers resistance against TEV but not other viruses in the potato virus Y group [31]; a PTGS transgene encoding the replicase protein of a European strain of PVX confers resistance against other European strains of PVX but not South Amer-

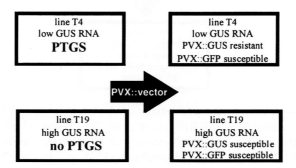

Fig. 3. Experimental testing of the relationship of PTGS and virus resistance. Tobacco line T4 exhibits a lower level of GUS expression than line T19 because the GUS transgene in line T4 exhibits PTGS. To find out whether PTGS in line T4 could confer virus resistance the plants were inoculated with PVX vectors. PVX::GUS carried a GUS reporter and had sequence homology to the silenced GUS transgene. PVX::GFP was not similar to the GUS transgene. The primary data were reported previously [14]

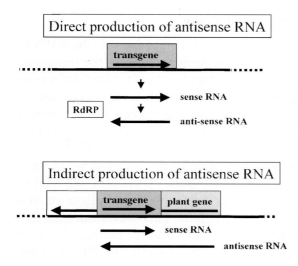

Fig. 4. Models for the production of antisense RNA in plants exhibiting PTGS. In the direct model the antisense RNA is produced by transcription from the transgene DNA. The promoter is a plant promoter located adjacent to the transgene locus. In the indirect model the antisense RNA is produced from the sense transgene RNA by an RNA dependent RNA polymerase (RdRP) encoded in the plant genome [44]

ican strains [32, 34]. Clearly the basis of this sequence specificity is central to understanding the mechanism of PTGS.

It seems inescapable that the PTGS specificity determinant is a nucleic acid [6]. Moreover, because the target gene in most examples of PTGS is a sense RNA, it is logical to propose that this nucleic acid is the antisense of the target RNA so that the specificity of the interaction would involve a direct interaction and Watson-Crick base pairing.

There is now a preliminary report of short antisense RNA in plants exhibiting PTGS [19] and our working hypothesis is that these antisense RNAs determine the specificity of the RNA targeting process. The previous failure to find these molecules could be related to their size. Most of the previous searches for antisense RNA were targeted at molecules in the range 200–5000bp in length and would not have detected RNA species that are very small.

We are exploring two models (Fig. 4) to explain how this antisense RNA could be produced. In the first, the transgene conferring PTGS integrates in the plant genome adjacent to an endogenous promoter [18]. According to this model the antisense RNA is produced by direct transcription from the endogenous promoter. The second model invokes indirect production of antisense RNA by a host encoded RNA-dependent RNA polymerase using the sense RNA of the silencing transgene as template [31]. There is no firm data to resolve either model. However the direct antisense transcription model seems unlikely because there is no correlation between the direct transcription of antisense RNA and PTGS [47]. Moreover, contrary to the prediction of the direct transcription model, the PTGS of a transgene was lost if the transgene promoter was suppressed [13]. On balance these findings are more compatible with the indirect formation of antisense RNA rather than the direct transcription model. However, the final answer about the production of antisense RNA will require targeted knockout of genes encoding the host encoded RNA polymerase so that the role of this enzyme in PTGS can be tested directly.

Signaling in gene silencing

I have referred above to the between-line variation in PTGS. Lines in which PTGS is highly active exhibit low-level transgene expression and (when the transgene is homologous to a virus) strain-specific resistance against virus infection. Conversely when the PTGS is weak the line exhibits high level expression of the transgene. To find out why there is this variation between lines it would be necessary to carry out a systematic analysis of transgene structures affecting PTGS. However this is a large scale and technically difficult undertaking.

In an attempt to short-cut the scale and difficulties of this analysis we investigated an alternative approach based on transient transgene expression. We refer to the experimental procedure as agroinfiltration because it involves infiltration of a leaf panel with a culture of *Agrobacterium tumefaciens*. This bacterium transfers part of its DNA into the genome of infected plant cells and we predicted that, if the transferred DNA had the necessary attributes, there would be PTGS in the infiltrated leaf panel.

To validate this approach we first carried out controls in which a GFP transgenic *Nicotiana benthamiana* was infiltrated with an *A. tumefaciens* culture carrying a conventional 35S:GFP construct [49, 50]. Surprisingly, even with these controls, there was PTGS in the infiltrated panel (Fig. 5). Even more surprisingly, the PTGS phenotype spread from the panel so that, by about 28d post infiltration, there was silencing of the GFP transgene throughout the plant [49, 50]. Like PTGS induced by integrated transgenes this effect was sequence specific: agroinfiltration with a GFP construct had no effect on a GUS transgene and, conversely, agroinfiltration with a GUS construct caused PTGS of a GUS transgene but had no effect on a GFP transgene. These findings led to the conclusion confirmed independently [37–39], that there is a systemic signal of gene silencing.

Systemic silencing probably plays a role in many examples of PTGS in transgenic plants. The initiating event may occur in just a single cell. Subsequently the signal molecule produced in that cell may spread to adjacent cells and initiate

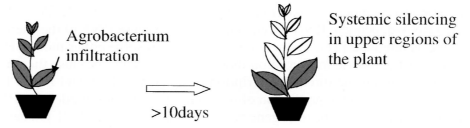

Fig. 5. Agroinfiltration to demonstrate systemic silencing of GFP. An agrobacterium culture carrying a 35S-GFP construct was infiltrated into the leaf of *N. benthamiana* expressing a 35S GFP construct at a high level. After 7-14d the GFP expression (shaded regions) was lost from systemic parts of the plant around the veins. Eventually the systemic silencing was complete in all regions of the plant except for the meristematic zones [50]

secondary signal production. Ultimately the degree, spatial distribution and kinetics of PTGS in the transgenic plant may be influenced by the production and systemic spread of the signal.

The identity of the signal molecule is currently unconfirmed. However, to explain the sequence specificity of the systemic silencing, it is difficult to avoid the conclusion that a nucleic acid is involved as the specificity determinant. One plausible scenario invokes antisense RNA, discussed above as the intracellular specificity determinant of PTGS, as the systemic signal.

Natural roles of PTGS

Several features of PTGS indicate that it is a transgenic manifestation of a natural virus resistance mechanism. First, in transgenic virus-resistant plants, viruses are effective targets of PTGS [5]. Second, viruses are able to induce PTGS. This virus induced gene silencing was first demonstrated in TEV infected lines carrying a TEV transgene [31]. As described above, the manifestation of induced gene silencing in these lines was a reduced level of the TEV transgene RNA and systemic resistance against TEV. Virus-induced gene silencing (VIGS) also takes place in non-transgenic plants when a virus vector carries host gene related sequences [28, 29, 43]. In this situation the symptoms on the infected plant depend on the function of the host gene. Thus, when the TMV or PVX vectors carried inserts that were based on the endogenous phytoene synthase, the photobleaching symptoms reflected the role of this enzyme in the biosynthesis of photoprotective carotenoid pigments [29, 43]. When the insert was from the gene for a chlorophyll biosynthetic enzyme there were yellow chlorotic symptoms [28]. Similarly an insert of a DNA polymerase caused arrest of growth and decrease in endopolyploidisation (Hamilton and Baulcombe, unpubl. obs.).

A third line of evidence that PTGS represents an antiviral defense is based on a series of cross protection experiments. Plants were first infected with a virus (the inducer) and subsequently with a second virus (the challenger). It is well established from this type of experiment that the inducer will only cross protect if the challenger is a related strain of virus. In our recent cross protection experiments with virus vectors we showed that cross protection was effective if the inducer and challenger virus were similar at the nucleotide level [41]. Provided that the two vectors carried similar insert sequences it was not necessary that they were taxonomically related or that the region of similarity was viral in origin. For example, a PVX vector was able to cross protect against a TMV vector provided that both constructs carried an insert from GFP [42]. Our interpretation of these data is that the cross protection is due to PTGS by the inducer. Consistent with this interpretation we showed that there is suppression of nuclear GFP reporter gene expression in tissue exhibiting cross protection provided that the PVX inducer virus carried a GFP-derived insert (Fig. 6).

Conclusive proof that PTGS is a defense mechanism can only be obtained by infection of plants that are mutated in genes required for PTGS [10]. If the model is correct these plants should be more susceptible to virus infection than the wild

Systemic accumulation of TMV::GFP and PVX::GUS

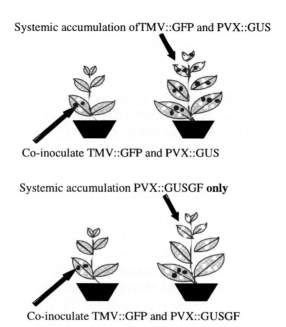

Co-inoculate TMV::GFP and PVX::GUS

Systemic accumulation PVX::GUSGF **only**

Co-inoculate TMV::GFP and PVX::GUSGF

Fig. 6. RNA sequence specificity in cross protection. *Nicotiana benthamiana* plants were inoculated with mixed TMV and PVX vector constructs. The TMV vector was TMV::GFP. The PVX vectors were either PVX::GUS or PVX::GUSGF in which the GF component represents the 5′ part of the GFP reporter gene. Cross protection between the two vector constructs was assessed in the systemically infected tissue [42]. In the plants inoculated with PVX::GUSGF and TMV::GFP there was an interaction involving the GFP sequence in the TMV construct and the GEP-derived GF in the PVX::GUSGF. The consequence of this interaction was cross protection and subsequently suppression of TMV::GFP in the systemic parts of the plant. We propose that this interaction involves a PTGS-like mechanism

type plants. However, in the absence of these data, there is strong indirect support for the virus resistance model from analyses of the Pl:hcpro protein encoded by potyviruses. These analyses follow from a series of elegant experiments showing that the virulence factor activity of this protein is due, at least in part, to suppression of a host defense mechanism [40]. Recently the earlier work has been tied in with PTGS by the demonstration in transgenic plants that P1:hcpro is also a suppressor of PTGS [2, 8, 27]. According to the model proposing that PTGS is host defense, the virulence factor and suppressor of PTGS activities would be different manifestations of the same process.

The ability of viral virulence factors to suppress PTGS is not unique to Pl:hcpro of potyviruses. The 2b protein encoded by cucumber mosaic virus also has these properties [8]. However the 2b protein does not suppress PTGS in the same way as P1:hcpro. Whereas the potyviral protein is able to reverse PTGS in all parts of the plant, the 2b protein is only able to prevent initiation of PTGS in tissues at the growing point of the plant [8]. These findings indicate that P1 : hcpro and 2b interfere with different components of the PTGS mechanism. It seems likely that many viruses will have adapted to PTGS by producing suppressors. It is also likely that viruses will have adopted alternative counter-PTGS strategies involving various types of evasion.

The existence of a silencing signal fits well with the proposal that PTGS represents a manifestation of antiviral defense [50]. A PTGS signal produced in virus-infected cells would have the potential to spread ahead of the infection front so that the non infected cells receiving the signal would be primed to activate PTGS targeted against the virus. Thus, the gene silencing-related resistance would have the potential to suppress virus spread as well as accumulation of the virus in infected cells.

However there is one feature of PTGS that is not easily reconciled with an antiviral defense role of PTGS. This feature is transgene DNA methylation. Transgenes conferring PTGS are ʹoften methylated in the transcribed region and there is strong evidence that the methylation is causal in the gene silencing mechanism [14, 23, 24, 45]. It is unlikely that DNA methylation concerns antiviral defense because most plant viruses have RNA rather than DNA genomes and do not produce DNA in their replication cycle. Therefore, it is likely that PTGS has several roles. It has been suggested that a mechanism related to PTGS may play a role in development [26]. However there is currently no strong evidence to support that idea and a second suggestion, that PTGS protects the genome against the effects of transposons [16], seems more plausible.

Gene silencing and functional genomics

It is likely that the complete arabidopsis genome will be sequenced in the next two years. Rice will be sequenced shortly afterwards and in the near future there will be extensive databases of expressed sequence tags from the genomes of many other crop plants. With the availability of these sequence data there is a pressing need for a high throughput technology that allows a function to be assigned to a genes of known sequence.

Virus-induced gene silencing (VIGS) will feature prominently in this functional genomics technology because it can be used to validate the role of sequenced genes. To exploit VIGS, a host-derived sequence is introduced into a PVX, TMV or geminivirus vector and a plant infected with this construct exhibits silencing of the corresponding gene. VIGS is effective with genes for phytoene metabolism, photosynthetic enzymes, proteins required for disease resistance and DNA replication ([28, 29, 43], and unpubl. data). We predict that most genes will be silenceable by VIGS. The few exceptions will probably include genes expressed exclusively in meristems because most viruses are not able to penetrate the meristematic zones of plants [43].

There are several advantages of VIGS over more conventional reverse genetics technologies based on targeted mutagenesis with transposons. For example, when the virus-based approach is targeted against essential genes, it may be possible to draw conclusions about the function of the target gene from the way that changes develop in the infected plant. In contrast, when the same gene is targeted by mutagenesis, the phenotype will often be embryo lethal and it will be difficult to extract interpretable information about the role of the gene.

A second situation in which virus-induced gene silencing will be preferred over mutagenesis is with multigene families. The effect of functional redundancy in the gene family will be to obscure the phenotype of a mutation. However, because gene silencing operates against RNA species with more than about 90% sequence identity, there can be a phenotype irrespective of the number of genes in the family.

Additional factors in favour of gene silencing as a functional genomics tool are speed and ease of use. To prepare the constructs is a routine molecular biology manipulation and takes no more than a few days. After inoculation the gene silencing phenotype is then observed within two to three weeks.

In future these procedures will be streamlined so that VIGS can be used as a forward as well as a reverse genetics tool. Most of the currently available virus vectors are in plasmids of *E. coli*. The vector constructs are assembled by ligation and then cloned in *E. coli*. The plasmid DNA is isolated from the clone, transcribed in vitro to produce infectious RNA and manually inoculated to a plant. However, if the virus vectors are in the expression cassette of an *A. tumefaciens* Ti plasmid vector [46], it will be possible to reduce the number of manipulations. There would be no advantage to these Ti plasmid vectors at the level of clone construction. However, once the agrobacterium clones have been prepared there is no need for further manipulations at the DNA level. The agrobacterium cells are applied to a wound site and infection follows. Presumably some of the cells at the wound site become transformed and serve as a source of virus inoculum for the rest of the plant.

In principle there is no practical reason why this agroinfection approach could not be used to test the function of thousands or even tens of thousands of different sequences. Many of the phenotypes in such an exercise will be obvious whereas others will be more cryptic and will require biochemical or other types of test on the infected plants. Disease resistance is one type of trait that may be particularly amenable to this type of approach: the plant will have compromised capability to mount a resistance response if an essential gene has been targeted by the virus vector. It is even possible that the approach could be used to target genes required for gene silencing. It would be a satisfying solution if enabling technology based on gene silencing could be used to identify the molecular components involved in the underlying mechanism.

Acknowledgement

I am grateful for support to The Sainsbury Laboratory from the Gatsby Charitable Foundation.

References

1. Abel PP, Nelson RS, De B, Hoffmann N, Rogers SG, Fraley RT, Beachy RN (1986) Delay of disease development in transgenic plants that express the tobacco mosaic virus coat protein gene. Science 232: 738–743

2. Anandalakshmi R, Pruss GJ, Ge X, Marathe R, Smith TH, Vance VB (1998) A viral suppressor of gene silencing in plants. Proc Natl Acad Sci USA 95: 13 079–13 084
3. Anderson JM, Palukaitis P, Zaitlin M (1992) A defective replicase gene induces resistance to cucumber mosaic virus in transgenic tobacco plants. Proc Natl Acad Sci USA 89: 8 759–8 763
4. Angenent GC, van den Ouweland JMW, Bol JF (1990) Susceptibility to virus infection of transgenic tobacco plants expressing structural and nonstructural genes of tobacco rattle virus. Virology 175: 191–198
5. Baulcombe DC (1996) Mechanisms of pathogen-derived resistance to viruses in transgenic plants. Plant Cell 8: 1 833–1 844
6. Baulcombe DC (1996) RNA as a target and an initiator of post-transcriptional gene silencing in transgenic plants. Plant Mol Biol 32: 79–88
7. Beck DL, Van Dolleweerd CJ, Lough TJ, Balmori E, Voot DM, Andersen MT, O'Brien IEW, Forster RLS (1994) Disruption of virus movement confers broad-spectrum resistance against systemic infection by plant viruses with a triple gene block. Proc Natl Acad Sci USA. 91: 10 310–10 314
8. Brigneti G, Voinnet O, Li WX, Ji LH, Ding SW, Baulcombe DC (1998) Viral pathogenicity determinants are suppressors of transgene silencing in Nicotiana benthamiana. EMBO J 17: 6 739–6 746
9. Cuozzo M, O'Connell KM, Kaniewski WK, Fang R-X, Chua N-H, Tumer NE (1998) Viral protection in transgenic tobacco plants expressing the cucumber mosaic virus coat protein or its antisense RNA. Bio/Technology 6: 549–557
10. Elmayan T, Balzergue S, Beon F, Bourdon V, Daubremet J, Guenet Y, Mourrain P, Palauqui JC, Vernhettes S, Vialle T, Wostrikoff K, Vaucheret H (1998) Arabidopsis mutants impaired in cosuppression. Plant Cell 10: 1 747–1 757
11. Elmayan T, Vaucheret H (1996) Expression of single copies of a strongly expressed 35S transgene can be silenced post-transcriptionally. Plant J 9: 787–797
12. English JJ, Baulcombe DC (1997) The influence of small changes in transgene transcription on homology-dependent virus resistance and gene silencing. Plant J 12: 1 311–1 318
13. English JJ, Davenport GF, Elmayan T, Vaucheret H, Baulcombe DC (1997) Requirement of sense transcription for homology-dependent virus resistance and trans-inactivation. Plant J 12: 597–603
14. English JJ, Mueller E, Baulcombe DC (1996) Suppression of virus accumulation in transgenic plants exhibiting silencing of nuclear genes. Plant Cell 8: 179–188
15. Fitchen JH, Beachy RN (1993) Genetically-engineered protection against viruses in transgenic plants. Annu Rev Microbiol 47: 739–763
16. Flavell RB (1994) Inactivation of gene expression in plants as a consequence of specific sequence duplication Proc Natl Acad Sci USA 91: 3 490–3 496
17. Golemboski DB, Lomonossoff GP, Zaitlin M (1990) Plants transformed with a tobacco mosaic virus nonstructural gene sequence are resistant to the virus. Proc Natl Acad Sci USA 87: 6 311–6 315
18. Grierson D, Fray RG, Hamilton AJ, Smith CJS, Watson CF (1991) Does co-suppression of sense genes in transgenic plants involve antisense RNA? Trends Biotechnol 9: 122–123
19. Hamilton A, Baulcombe DC (1998) Investigations into the use and mechanisms of gene silencing. In: International Congress of Plant Pathology, Edinburgh, pp 1.12.20
20. Hemenway CL, Fang RX, Kaniewski WK, Chau NH, Tumer NE (1988) Analysis of the mechanism of protection in transgenic plants expressing the potato virus X coat protein or its antisense RNA. EMBO J 7: 1 273–1 280

21. Hobbs SLA, Kpodar P DeLong CMO (1990) The effect of T-DNA copy number, position and methylation on reporter gene expression in tobacco transformants. Plant Mol Biol 15: 851–864

22. Hobbs SLA, Warkentin TD, DeLong CMO (1993) Transgene copy number can be positively or negatively associated with transgene expression, Plant Mol Biol 21: 17–26

23. Ingelbrecht I, Van Houdt H, Van Montagu M, Depicker A (1994) Posttranscriptional silencing of reporter transgenes in tobacco correlates with DNA methylation. Proc Natl Acad Sci USA 91: 10 502–10 506

24. Jones AL, Thomas CL, Maule AJ (1998) De novo methylation and co-suppression induced by a cytoplasmically replicating plant RNA virus. EMBO J 17: 6 385–6 393

25. Jorgensen RA (1990) Altered gene expression in plant due to trans interactions between homologous genes. Biotechnol Trends 8: 340–344

26. Jorgensen RA, Atkinson RG, Forster RLS, Lucas WJ (1998) An RNA-based information superhighway in plants. Science 279: 1 486–1 487

27. Kasschau KD, Carrington JC (1998) A counterdefensive strategy of plant viruses: suppression of post-transcriptional gene silencing. Cell 95: 461–470

28. Kjemtrup S, Sampson KS, Peele CG, Nguyen LV, Conkling MA, Thompson WF, Robertson D (1998) Gene silencing from plant DNA carried by a geminivirus. Plant J 14: 91–100

29. Kumagai MH, Donson J, Della-Cioppa G, Harvey D, Hanley K, Grill LK (1995) Cytoplasmic inhibition of carotenoid biosynthesis with virus-derived RNA. Proc Natl Acad Sci USA 92: 1 679–1 683

30. Lawson C, Kaniewski WK, Haley L, Rozman R, Newell C, Sanders P, Tumer NE (1990) Engineering resistance to mixed virus infection in a commercial potato cultivar: resistance to potato virus X and potato virus Y in transgenic Russet Burbank. Bio/Technology 8: 127–134

31. Lindbo JA, Silva-Rosales L, Proebsting WM, Dougherty WG (1993) Induction of a highly specific antiviral state in transgenic plants: implications for regulation of gene expression and virus resistance. Plant Cell 5: 1 749–1 759

32. Longstaff M, Brigneti G, Boccard F, Chapman SN, Baulcombe DC (1993) Extreme resistance to potato virus X infection in plants expressing a modified component of the putative viral replicase. EMBO J 12: 379–386

33. Matzke MA, Primig M, Trnovsky J, Matzke AJM (1989) Reversible methylation and inactivation of marker genes in sequentially transformed tobacco plants. EMBO 8: 643–649

34. Mueller E, Gilbert JE, Davenport G, Brigneti G, Baulcombe DC (1995) Homology-dependent resistance: transgenic virus resistance in plants related to homology-dependent gene silencing. Plant J 7: 1 001–1 013

35. Napoli C, Lemieux C, Jorgensen RA (1990) Introduction of a chimeric chalcone synthase gene into Petunia results in reversible co-suppression of homologous genes in trans. Plant Cell 2: 279–289

36. Osbourn JK, Watts JW, Beachy RN, Wilson TMA (1989) Evidence that nucleocapside disassembly and a later step in virus replication are inhibited in transgenic tobacco protoplasts expressing TMV coat protein. Virology 172: 370–373

37. Palauqui J-C, Balzergue S (1999) Activation of systematic acquired silencing by localised introduction of DNA Curr Biol 9 (in press)

38. Palauqui J-C, Elmayan T, Pollien J-M, Vaucheret H (1997) Systemic acquired silencing: transgene-specific post-transcriptional silencing is transmitted by grafting from silenced stocks to non-silenced scions. EMBO J 16: 4 738–4 745

39. Palauqui JC, Vaucheret H (1998) Transgenes are dispensable for the RNA degradation step of cosuppression. Proc Natl Acad Sci USA: 9 675–9 680

40. Pruss G, Ge X, Shi XM, Carrington JC, Vance VB (1997) Plant viral synergism: the potyviral genome encodes a broad-range pathogenicity enhancer that transactivates replication of heterologous viruses. Plant Cell 9: 859–868

41. Ratcliff F, Harrison BD, Baulcombe DC (1997) A similarity between viral defense and gene silencing in plants. Science 276: 1 558–1 560

42. Ratcliff F, MacFarlane S, Baulcombe DC (1999) Gene silencing without DNA: RNA-mediated cross protection between viruses. Plant Cell (in press)

43. Ruiz MT, Voinnet O, Baulcombe DC (1998) Initiation and maintenance of virus-induced gene silencing. Plant Cell 10: 937–946

44. Schiebel W, Pelissier T, Reidel L, Thalmeir S, Schiebel R, Kempe D, Lottspeich F, Sanger HL, Wassenegger M (1998) Isolation of an RNA-directed RNA polymerase-specific cDNA clone from tomato. Plant Cell 10: 2 087–2 102

45. Sijen T, Wellink J, Hiriart J-B, van Kammen A (1996) RNA-mediated virus resistance: role of repeated transgenes and delineation of targeted regions. Plant Cell 8: 2 277–2 294

46. Turpen TH, Turpen AM, Weinzettl N, Kumagai MH, Dawson WO (1993) Transfection of whole plants from wounds inoculated with Agrobacterium tumefaciens containing cDNA of tobacco mosaic virus. J Virol Methods 42: 227–240

47. Van Blokland R, Van der Geest N, Mol JNM, Kooter JM (1994) Transgene-mediated suppression of chalcone synthase expression in Petunia hybrida results from an increase in RNA turnover. Plant J 6: 861–877

48. van Dun CMP, Bol JF, Van Vloten-Doting L (1987) Expression of alfalfa mosaic virus and tobacco rattle virus coat protein genes in transgenic tobacco plants. Virology 159: 299–305

49. Voinnet O, Baulcombe DC (1997) Systemic signalling in gene silencing. Nature 389: 553

50. Voinnet O, Vain P, Angell S, Baulcombe DC (1998) Systemic spread of sequence-specific transgene RNA degradation is initiated by localised introduction of ectopic promoterless DNA. Cell 95: 177–187.

Authors' address: Prof. D. Baulcombe, The Sainsbury Laboratory, John Innes Centre, Norwich NR4 7UH, U.K.

Viroids and the nature of viroid diseases*

T. O. Diener

Center for Agricultural Biotechnology and Department of Molecular Genetics
and Microbiology, University of Maryland, College Park, Maryland, U.S.A.

Summary. In its methodology, the unexpected discovery of the *viroid* in 1971 resembles that of the virus by Beijerinck some 70 years earlier. In either case, a novel type of plant pathogen was recognized by its ability to penetrate through a medium with pores small enough to exclude even the smallest previously known pathogen: bacteria as compared with the tobacco mosaic agent; viruses as compared with the potato spindle tuber agent. Interestingly, one of the two methods used by Beijerinck, diffusion of the tobacco mosaic agent into agar gels, is conceptually similar to one method used to establish the size of the potato spindle tuber agent, namely polyacrylamide gel electrophoresis. Further work demonstrated that neither agent is an unusually small conventional pathogen (a microbe in the case of the tobacco mosaic agent; a virus in the case of the potato spindle tuber agent), but that either agent represents the prototype of a fundamentally distinct class of pathogen, the viruses and the viroids, respectively.

With the viroids, this distinction became evident once their unique molecular structure, lack of mRNA activity, and autonomous replication had become elucidated. Functionally, viroids rely to a far greater extent than viruses on their host's biosynthetic systems: Whereas translation of viral genetic information is essential for virus replication, viroids are totally dependent on their hosts' transcriptional system and, in contrast to viruses, no viroid-coded proteins are involved.

Because of the viroids' simplicity and extremely small size they approach more closely even than viruses Beijerinck's concept of a *contagium vivum fluidum*.

Introduction

Recognition of the fundamental disparity between viruses and viroids became possible only after certain basic principles of 'virology and molecular biology had been established. These principles helped create an intellectual climate in

*Portions of this presentation are expanded and modified versions of a paper entitled "Portraits of Viruses: the Viroid" [Intervirology 22: 1–16 (1984)], published by S. Karger A. G., Basel, Switzerland

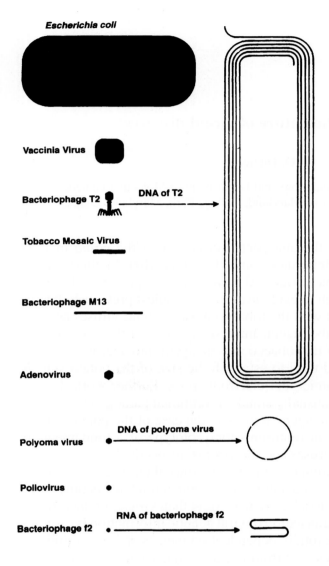

Fig. 1. Size comparison between a bacterium, several viruses, and the viroid. Adapted from Scientific American 244: 66–73 (1981)

which the existence of free nucleic acid pathogens could not *a priori* be ruled out. Aside from general biological principles, at least four important prerequisites can be identified: (i) Foremost was the realization that the genetic information of viruses resides in their nucleic acid component, a fact that, in the case of plant viruses, was most dramatically established with the demonstration that RNA isolated from tobacco mosaic virus is infectious [35, 37]; (ii) the discovery of the so-called NM forms of tobacco rattle virus and the demonstration that these consist of unencapsidated viral RNA [74] and that viral pathogens may exist in nature in the form of free RNA [7]; (iii) the isolation of defective tobacco mosaic virus strains from nitrous acid-treated virus preparations and the demonstration that with these the infectious principle behaves in a manner similar to RNA isolated from ordinary virus [84], clearly showing that a virus may be able not only to persist in vivo in the form of free RNA but to spread from cell to cell; and (iv) with the demonstration that free, infectious RNA exists also in plants infected with a conventional plant virus [12], the possibility that viruses existing only as free RNA might occur in nature became still more plausible. Discovery of a free nucleic acid pathogen would therefore not have been too surprising – the scientific mind was prepared for it. What it was clearly not prepared to accept, however, was a pathogen with the characteristics of the viroid which conflicted with tenets widely held by molecular biologists and virologists (Fig. 1).

Viroid discovery

In the 1960s, several investigators attempted to purify the virus thought to be responsible for the potato spindle tuber disease (Fig. 2), but none of these efforts was successful.

Some investigators discovered, however, that the putative virus would not readily sediment in an ultracentrifuge; still others found that the infectious principle in nucleic acid extracts from infected plants migrated in polyacrylamide

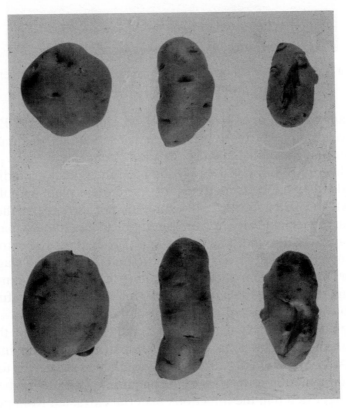

Fig. 2. Tuber symptoms of potato spindle tuber disease in *Solanum tuberosum*. Upper row, cv. Saco; lower row, cv. Kennebec. Left: healthy; center: infected with the type strain of PSTVd; right: infected with a more severe strain

Fig. 3. Symptoms of PSTVd infection in *Lycopersicon esculentum*, cv. Rutgers. Left: healthy
plant; right: infected plant

gels faster than expected of a 1- or 2-million M_r nucleic acid (R. P. Singh, pers. comm.). However, none of these preliminary observations was reported or followed up prior to the establishment of the viroid concept.

Early in the 1960s, efforts to purify the presumed potato spindle tuber virus were also made by two U.S. Department of Agriculture plant pathologists, W. B. Raymer and M. J. O'Brien. In 1962, these investigator showed that the potato spindle tuber agent is mechanically transmissible to Rutgers tomato plants, in which it causes a characteristic syndrome [69] (Fig. 3). They prepared extracts with relatively high infectivity titers from potato or tomato leaves. The stage, therefore, seemed set for the purification of the presumed virus, but when Raymer attempted to pellet the agent, he found that, even after prolonged ultracentrifugation, most of the infectious principle remained in the supernatant solution.

In collaboration with Raymer, the low sedimentation rate of the infectious agent was confirmed by rate-zonal centrifugation of extracts from infected plants [27] (Fig. 4). Also, treatment of extracts with phenol was shown not to appreciably change the sedimentation properties of the agent [27]. Results of these and other experiments suggested that the infectious material extracted from potato spindle tuber-affected tissue was a free nucleic acid and not a conventional viral nucleoprotein particle. Because incubation with ribonuclease A, but not incubation with deoxyribonuclease or pronase, abolished infectivity, it became evident that

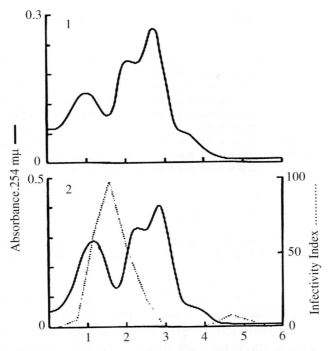

Fig. 4. Distribution of UV absorbance and infectivity in centrifuged sucrose density-gradient columns containing extracts from: *1* healthy and *2* infected tissue. Centrifugation for 16 h at 50,000 **g**. Note lack of UV-absorbing component coinciding with infectivity distribution

RNA was an essential portion, if not all, of the infectious agent and that proteins probably were not involved. By 1967, it had become evident that the infectious principle in the extracts from infected leaves was free RNA and that no virus-like particles were present [27].

These results raised a number of puzzling questions. Did the infectious RNA originate from very unstable virions that did not survive the mild extraction procedure used? Vacuum infiltration of RNase into infected leaves and other indications showed that this was not the case [14]. Was the RNA single- or double-stranded? Was it as small as suggested by its low rate of sedimentation? Resolution of these questions was difficult because the RNA could not be recognized as a physical entity, such as a peak in a UV-light absorption profile of a centrifuged sucrose gradient tube or a band in a polyacrylamide gel. Evidently, the RNA was present in very small amounts. Position of the RNA could only be deduced by virtue of its biological activity, that is, by inoculation of tomato plants with all gradient fractions or extracts from gel slices. Progress was further handicapped because no local lesion assay was available and because titer estimates based on systemic symptoms are inherently inaccurate [3]. Nevertheless, if properly done, assays of gel slices or density gradient fractions on tomato plants permitted locating the position of the infectious agent rather accurately (Fig. 5).

Crude and laborious as these methods were, they revealed all essential properties that distinguish viroids from viruses. Thus, the very low M_r of the infectious

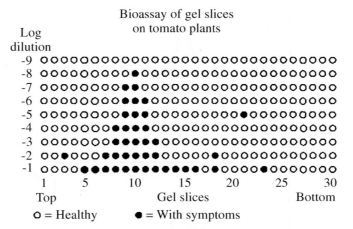

Fig. 5. Actual tomato bioassay results of consecutive slices from electrophoresed, 20% polyacrylamide gel. Left: slices from top; right: slices from bottom of gel

RNA could be established unequivocally by comparing the results of sedimentation analysis and electrophoretic analyses – all achieved solely by appropriate bioassays on tomato plants [15]. Although initial molecular size estimates were too low, we now know the reason for this discrepancy: the highly base-paired, compact native structure of the viroid permits it to migrate faster in gels than do RNA markers of equal M_r but more usual structure.

The small size of the RNA suggested that it might be akin to a satellite RNA, requiring a helper virus for its own replicaton, but extensive experimentation failed to substantiate this contention [15]. This left only one conceivable possibility of accommodating the pathogen within accepted tenets of virology: The RNA might consist of a population of small RNA molecules that, *in toto*, could represent the equivalent of a viral genome of more or less conventional size. In 1971–1972, however, first indications were reported that this did not appear to be the case and that the RNA most likely was a single molecular species [13, 16, 17].

Thus, by 1971, it had become evident that the potato spindle tuber agent was not simply an unusual virus, but represented the prototype of a new class of pathogen with physical/chemical properties fundamentally different from those of known viruses in at least four important respects: (i) the pathogen exists in vivo as an unencapsidated nucleic acid, i.e., no virus-like particles are present in infected tissue; (ii) the pathogen is a low-molecular-weight RNA; (iii) the infectious RNA, despite its small size, is replicated autonomously in susceptible cells, that is, without requiring a helper virus; and (iv) the infectious RNA consists of one molecular species only. Recognition of the unique properties of the potato spindle tuber agent led to the proposal to call it and similar agents *viroids* [15].

Confirmation

Indications that the potato spindle tuber viroid (PSTVd) was not the only plant pathogen of its kind had become known as early as 1968. Thus, in crude extracts, properties, of the chrysanthemum stunt [60] and citrus exocortis [80, 81] agents were shown to resemble those of the potato spindle tuber agent reported earlier [27]. Indeed, the viroid nature of the chrysanthemum stunt and citrus exocortis agents could readily be demonstrated [26, 74, 82]. This rapid and independent confirmation of the original SPTVd work with two other plant pathogens and involving two other laboratories undoubtedly was an important factor in the ready acceptance of the viroid concept among plant virologists.

Doubts and final acceptance

Outside of plant virology, however, acceptance of the viroid concept was slow. It must be remembered that, at the time, all properties of viroids had to be determined by bioassay on indicator plants and not by conventional biophysical/biochemical techniques. Many molecular biologists and molecularly oriented virologists considered evidence obtained in this fashion unconvincing, if not altogether unacceptable. The fact that the viroid concept clashed with the widely held, but mostly unspoken belief that an autonomously replicating viral entity required genetic information equivalent to a minimum of $(0.9-1.0) \times 10^6 M_r$ of nucleic acid (values derived from the small RNA phages) did not help the situation. For some time the scientific community-at-large regarded the viroid with a healthy dose of skepticism or adopted a wait-and-see attitude. What was clearly needed was to purify the viroid and to determine its physical/chemical properties by conventional methods.

Purification

Unequivocal recognition of a viroid as a physical entity was first achieved with PSTVd [18] (Fig. 6), and relatively pure preparations soon became available [11, 18, 76, 79]. Although, for a time, purification of viroids remained a formidable and time consuming nucleic acid separation problem, it did become feasible and permitted, for the first time, application of routine biophysical/biochemical analyses to the viroid. This led to rapid advances in our knowledge of viroid structure. The following milestones might be mentioned: The thermal denaturation properties of a viroid (PSTVd) were determined and shown to be intermediate to those of genuine single- or double-stranded RNAs [18], explaining the earlier difficulties in deciding the strandedness of the RNA; Sogo et al. [85] achieved visualization of native PSTVd molecules by electron microscopy and confirmed, by direct length measurement, the low molecular weight of the viroid, they concluded that native PSTVd is a hairpin-like, highly base-paired, single-stranded RNA; Semancik and colleagues determined the nucleotide composition of the citrus exocortis viroid (CEVd) and its thermal denaturation properties [79], which were found to be much like those of PSTVd; Dickson et al. [11] obtained the first RNA fingerprints of viroids and demonstrated that PSTVd and CEVd are distinct and different species of low-molecular-weight RNA; Sänger et al. [77] and McClements and Kaesberg [62] succeeded in visualizing partially and completely denatured viroids by electron microscopy and made the important discovery that

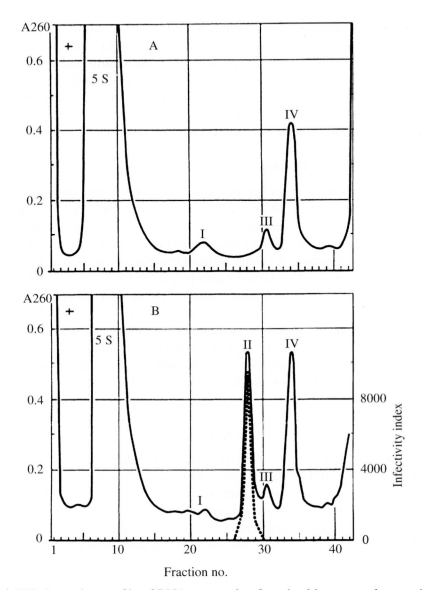

Fig. 6. A UV-absorption profile of RNA preparation from healthy tomato leaves after electrophoresis in a 20% polyacrylamide gel for 7.5 h at 4 °C. **B** UV-absorption (————) and infectivity distribution (– – – –) profiles of RNA preparation from PSTVd-infected tomato leaves after electrophoresis as in **A**. 5S, 5S ribosomal RNA; *I, III, IV* unidentified minor components of cellular RNA; *II* PSTVd. Electrophoretic movement from right to left

many, if not most, viroid molecules have a covalently closed circular structure, the first such RNAs found in nature; Owens et al. [66] provided evidence that both circular and linear PSTVd molecules are infectious; intensive biophysical and biochemical studies of viroids [30, 41, 48, 59] were crowned by the determination of the complete nucleotide sequence and probable secondary structure of PSTVd by Gross et al. [40] (Fig. 7). PSTVd thus became the very first pathogen

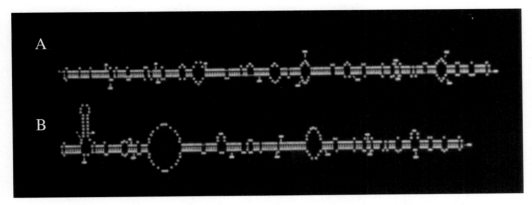

Fig. 7. Alternative computer-generated secondary structures of PSTVd

of a eukaryotic organism for which the complete molecular structure had been established.

Later work has confirmed these results and has shown that almost all viroids can assume the characteristic, thermodynamically most stable, highly base-paired, rodlike secondary structure (the "native" conformation) first described for PSTVd [77], in which short base-paired regions alternate with mismatched internal and bulge loops. Also generally applicable so far are the five distinct topological/functional domains first proposed by Keese and Symons [57]: the two terminal domains, a "pathogenicity" domain, a central domain, and a variable domain. A more detailed description of viroid molecular structure is beyond the scope of this paper; the reader is referred to excellent reviews [47, 70, 71].

It is worth noting that, in contrast to many other biological systems, viroids are unique in that our knowledge of their molecular properties far outstrips our knowledge of their basic biological characteristics, such as mode of transmission in nature, natural host range, and epidemiological parameters. Thus, although viroids have reached maturity in one area, they are still far from fully understood in others.

Additional viroids

In parallel with the intensive work on the physical/chemical properties of PSTVd, more biologically oriented work resulted in the discovery of many additional members of the viroid family [1, 2, 22, 33, 38, 46, 49, 61, 67, 68].

The primary sequences and most probable secondary structures of more than two dozen viroid species and of many more viroid variants have been published (for a recent compilation, see [33]).

Viroid function

Elucidation of viroid function has been one of the most intriguing and frustrating challenges. How and why is this bit of genetic information replicated in a susceptible cell? How does the viroid interfere with the host cells' metabolism to produce, in some plants at least, the diseases that brought the viroid to our attention in the first place?

Prima facie, the early finding that actinomycin D inhibits viroid replication [28, 63] seemed to suggest that viroids might be replicated from DNA templates, but with the identification of viroid-complementary RNA sequences in infected tissue [39] and the demonstration that these viroid complements are of full size (and therefore suitable as templates) [65], involvement of DNA in viroid replication became unlikely.

Today, general consensus exists that viroids are (i) not translated into viroid-specified polypeptides [10, 45], (ii) replicated by host enzymes, probably normally DNA-dependent RNA polymerase II, from RNA templates [78], and (iii) probably replicated by a rolling circle-type mechanism with the circular viroid serving as template, resulting in the synthesis of oligomeric strands of the viroid complement [5] (Fig. 8).

Oligomers are subsequently cleaved into monomers and ligated to form the circular progeny viroids. Of necessity, this cleavage must be precise. Two different processes appear to be operative: With three viroids (the avocado sunblotch viroid [53], peach latent mosaic viroid [50], and chrysanthemum chlorotic mottle viroid [64]), cleavage apparently occurs nonenzymatically by means of conserved sequences that can base-pair into what has become known as "hammerhead" structures, in which configuration cleavage occurs at a precise position in the molecule. With other viroids, the cleavage mechanism is unknown, except that specific host ribonucleases may be involved [86].

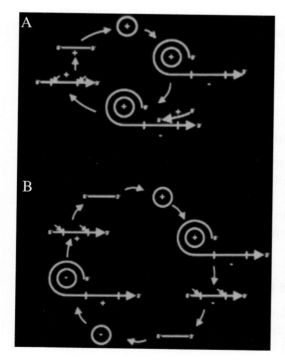

Fig. 8. Alternative schemes of rolling circle replication of viroids. **A** Asymmetric transcription, in which nonomeric circular plus strands serve as templates for the transcription (presumably by normally DNA-directed host RNA polymerase II, see text) of oligomeric linear minus strands, which are cleaved into monomeric plus strands, and the latter ligated to result in circular, plus strand progeny viroids. **B** Symmetric transcription, in which oligomeric minus strands are cleaved into minus monomers which, after ligation, serve as rolling circle templates for the production of oligomeric plus strands

Viroid pathogenicity

The lack of mRNA activity of viroids implies that all detrimental effects on the host plant of viroid infection must be the result of direct specific interaction of the viroid (or its complement) with so far undetermined host constituents [15, 34]; viroids may thus be thought of as abnormal constituents of host metabolism [20]. Comparisons of viroid nucleotide sequences with those of cellular RNAs have revealed a number of similarities. Thus, a portion of the negative strand of PSTVd is complementary with the 5′-terminus of U1 snRNA [21] and 7S RNA from tomato leaf tissue displays notable sequence similarity with part of the pathogenicity domain of PSTVd [42]; and it has been suggested that complex formation between the viroid and U1 or 7S RNA could interfere with pre rRNA processing or with formation of signal recognition particles, respectively.

These hypotheses cannot explain, however, why sometimes infection of one species by a viroid results in severe symptoms, whereas infection of another, often closely related, species with the same viroid does not lead to the induction of detectable symptoms; this despite the fact that the viroid titer reached in the symptom-bearing species often does not significantly differ from that in the symptomless species [19]. Also, some mild and severe strains of PSTVd reach similar titers in tomato, but induce symptoms of drastically different severity.

Clearly, to be plausible, any hypothesis purporting to explain the mechanisms of viroid pathogenesis must be based on one or more host metabolic systems capable of responding differentially to infection by viroid strains of varying pathogenicity. One possibility is that individual viroid strains differentially bind to certain host proteins and that the strength of this binding affects the severity of symptom formation. Indeed, a 43 kD host protein (as well as histones) binds to PSTVd when the viroid is mixed in vitro with nuclear extracts, but whether such complexes have any biological significance has not been determined [58, 88].

In another study, a host-encoded 68 kD protein appeared to be more highly activated (phosphorylated) in extracts from PSTVd- infected, as compared to extracts form mock-infected tissue [51]. Preliminary evidence suggested that the protein was a dsRNA-dependent protein kinase immunologically related to a mammalian interferon-induced, dsRNA-activated, M_r 68 K protein kinase (PKR or p68). PKR has been implicated in the regulation of virus synthesis [55]. The enzyme is characterized by two distinct kinase activities: (i) autophosphorylation (activation) and (ii) kinase activity on exogenous substrates. Activated PKR phosphorylates, via a cascade of reactions, its natural substrate, the alpha subunit of eukaryotic protein synthesis initiation factor eIF-2 [52]. These reactions lead to reductions in functional eIF-2, culminating in the impediment of protein synthesis initiation [54]. Several investigators have documented that virus-specific RNAs synthesized during infection have the potential to activate PKR [56, 72].

Whether viroids trigger similar reactions involving a plant equivalent of PKR is unknown but, in analogy with the results with animal viruses, appears possible. Involvement of such an enzyme in viroid pathogenicity was further suggested by the demonstration that purified PSTVd, when added to an in vitro assay system containing purified mammalian PKR, activated the enzyme and that RNA transcripts of PSTVd cDNAs specifically bound to a PKR-antibody complex [25]. Activation of PKR by the RNA of a strain of PSTVd that results in severe symptoms in tomato plants was at least ten-fold that induced by the RNA of a mild strain. These results represent the first demonstration of a differential effect of viroid strains inducing different levels of pathology on any biochemical or metabolic host systems investigated [25].

Recently, a protein of ca. 68kD isolated from tomato has been identified as a plant equivalent of mammalian dsRNA-activated protein kinase on the basis of its reaction with monoclonal and polyclonal antibodies specific to human PKR and its characteristic autophosphorylation in the presence of dsRNAs [89]. In vitro RNA transcripts, equivalent to (+)- strand, monomeric PSTVd, specifically bind to the tomato enzyme. Experiments with partial PSTVd transcripts showed that binding occurs to the pathogenicity domain-containing left half of PSTVd, but not to the right half [89], strengthening earlier indication that binding of viroids to host PKR may play an important role in viroid pathogenesis.

Origin of crop viroids

Years ago, I speculated that viroid diseases of crop plants may originate by chance transfer of viroids from reservoirs in wild plants [19]. My hypothesis was based on the observation that all known crop viroid diseases appear to be of recent origin (20th century) and that several of them abruptly appeared seemingly *de novo*. With PSTVd, however, no wild-plant reservoirs have been found, either in the United States (where the disease was first observed) or in the Andes of South America (the original location of the potato). Thus the origin of PSTVd has long remained a mystery.

Recently, however, a novel viroid, the Mexican papita viroid (MPVd) has been discovered in Mexico, where it harmlessly inhabits a tuber-bearing, wild solanaceous plant, *Solanum cardiophyllum* [61]. Evolutionarily, MPVd is most closely related to TPMVd and PSTVd. It is known that specimens of *S. cardiophyllum* have been introduced to the United States in the latter part of the 19th century in efforts to find resistance to the late blight fungus (*Phytophthora infestans*). It is possible, therefore, that some of these introduced plants were endemically infected with MPVd or a similar viroid and that the potato spindle tuber disease originated by chance transfer of the viroid from such infected germplasm plants.

The question of animal viroids

All known viroids are pathogens of higher plants and, at present, the only connection between plant viroids and animal agents is the hepatitis delta virus (HDV) RNA [44, 87] which appears to be the product of recombination between a viral and a viroidlike RNA – with the latter exhibitiing significant sequence similarities with viroids [4]. Phylogenetically, the viroidlike portion of HDV RNA appears to be related to plant viroids [31].

Viroid-like RNAs and DNAs

Although viroids were the first small, circular, RNAs to be discovered, a whole universe of small, mostly circular, RNAs has since come to light. Such RNAs from plants include encapsidated viroid-like satellite RNAs [36], certain other plant satellite RNAs, such as the satellite RNA of tobacco ringspot virus [6]; VS RNA and DNA from mitochondria of certain *Neurospora* isolates [73], and a novel retroviroid-like element from carnation (carnation small viroid-like RNA and DNA [9]). From animal sources, they include the RNA of hepatitis delta virus [44], and a small transcript from a newt satellite DNA [32].

A detailed description of these RNAs is beyond the scope of this article. Suffice it to say that all of these RNAs (and DNAs) are related by virtue of the fact that they possess certain, but not all, characteristics of viroids. Thus, most replicate via RNA intermediates by rolling-circle mechanisms, but others are transcribed from DNA templates; most exist as covalently closed, circular molecules, but some occur mostly as linear molecules; except for some satellite RNAs and hepatitis delta virus RNA, none possesses functional open reading frames; some replicate autonomously, others require a helper virus; except for VS RNA, none exhibits significant nucleotide similarities to either host DNA or to helper virus genomes; oligomeric replication intermediates of some self-cleave (often by "hammerhead" structures), whereas others do not self-cleave.

Viroid evolution

Phylogenetic analysis suggests that viroids and viroidlike satellite RNAs, as well as the viroidlike domain of hepatitis delta virus RNA, constitute a monophyletic group [31], but whether any of the other small RNAs are evolutionarily related to viroids is unknown. Viroids, viroidlike satellite RNAs and some of the other small, circular RNAs display significant sequence similarties with introns, particularly with the conserved sequences of group I introns [24, 29, 43]; and, in analogy with speculations concerning intron evolution [8, 83], it has been suggested that viroids may represent "living fossils" of the hypothetical RNA world [23, 24].

Epilogue

Many years passed before the object of Beijerinck's landmark discovery, the "*contagium vivum fluidum*," was stripped of its mysterious aura and the virus concept established on a sound physical/chemical and biological foundation.

In contrast, the viroid concept, which may be regarded as a 50-fold extension of Beijerinck's original *contagium vivum fluidum* into still smaller dimensions, was never mysterious from a physical/chemical standpoint. Very early, it had become evident that the viroid consists of RNA and of RNA alone. On the other hand, with respect to its biological properties, mystery has not altogether been banished yet. Still, no convincing answers can be given to many question. Why, for example, do viroids occur apparently only in higher plants and not in animals; why, as free RNAs, are viroids so contagious under field conditions, and why is the viroid replicated when introduced into susceptible cells, whereas cellular

RNAs, similarly introduced, are broken down? Evidently, many gaps still exist in our knowledge of viroids and viroid diseases.

Present knowledge is sufficient, however, to regard viroids as representing the most extreme form of parasitism known. Viroids depend for their replication to a far greater degree even than viruses on their hosts' metabolic machinery. This property, in addition to their extreme smallness, qualifies viroids as the ultimate *contagium vivum fluidum* in Beijerinck's terminology.

References

 1. Ambrós S, Desvignes JC, Llácer G, Flores R (1995) Pear blister canker viroid: sequence variability and causal role in pear blister canker disease. J Gen Virol 76: 2 625–2 629
 2. Ashulin L, Lachman O, Hadas R, Bar-Joseph M (1991) Nucleotide sequence of a new viroid species, citrus bent leaf viroid (CBLVd) isolated from grapefruit in Israel. Nucleic Acids Res 19: 4 767
 3. Brakke MK (1970) Systemic infections for the assay of plant viruses. Ann Rev Phytopathol 8: 61–84
 4. Branch AD, Levine BJ, Robertson HD (1990) The brotherhood of circular RNA pathogens: Viroids, circular satellites, and the delta agent. Semin Virol 1: 143–152
 5. Branch AD, Robertson HD (1984) A replication cycle for viroids and other small infectious RNA's. Science 223: 450–455
 6. Bruening G, Passmore BK, van Tol H, Buzayan JM, Feldstein PA (1991) Replication of a plant virus satellite RNA: evidence favors transcription of circular templates of both polarities. Mol Plant-Microbe Interact 4: 219–225
 7. Cadman CH (1962) Evidence for association of tobacco rattle virus nucleic acid with a cell component. Nature 193: 49–52
 8. Cech TR (1986) A model for the RNA-catalyzed replication of RNA. Proc Natl Acad Sci USA 83: 4 360–4 363
 9. Darós JA Flores R (1995) Identification of a retroviroid-like element from plants. Proc Natl Acad Sci USA 92: 6 856–6 860
10. Davies JW, Kaesberg P, Diener TO (1974) Potato spindle tuber viroid. XII. An investigation of viroid RNA as a messenger for protein synthesis. Virology 61: 281–286
11. Dickson E, Prensky W, Robertson HD (1975) Comparative studies of two viroids: analysis of potato spindle tuber and citrus exocortis viroids by RNA fingerprinting and polyacrylamide-gel electrophoresis. Virology 68: 309–316
12. Diener TO (1962) Isolation of an infectious, ribonuclease-sensitive fraction from tobacco leaves recently inoculated with tobacco mosaic virus. Virology 16: 140–146
13. Diener TO (1970) Potato spindle tuber virus: a plant virus with properties of a free ribonucleic acid. Xth International Congress Microbiology, p 238
14. Diener TO (1971) Potato spindle tuber virus: A plant virus with properties of a free nucleic acid. III. Subcellular location of PSTV-RNA and the question of whether virions exist in extracts or in situ. Virology 43: 75–89
15. Diener TO (1971) Potato spindle tuber "virus". IV. A replicating, low molecular weight RNA. Virology 45: 411–428
16. Diener TO (1972) Viroids. Adv Virus Res 17: 295–313
17. Diener TO (1972) Potato spindle tuber virus, a replicative low molecular weight RNA. In: Melnick JL (ed) International Virology 2. Budapest, Hungary, Second International Congress Virology, Budapest, 1971, p 281
18. Diener TO (1972) Potato spindle tuber viroid. VIII. Correlation of infectivity with a

UV-absorbing component and thermal denaturation properties of the RNA. Virology 50: 606–609

19. Diener TO (1979) Viroids and viroid diseases. Wiley, New York, 273pp.

20. Diener TO (1981) Viroids: abnormal products of plant metabolism. Ann Rev Plant Physiol 32: 313–25

21. Diener TO (1981) Are viroids escaped introns? Proc Natl Acad Sci USA 78: 5 014–5 015

22. Diener TO (ed) (1987) The viroids. Plenum, New York

23. Diener TO (1989) Circular RNAs: relics of precellular evolution? Proc Natl Acad Sci USA 86: 9 370–9 374

24. Diener TO (1996) Origin and evolution of viroids and viroidlike satellite RNAs. Virus Genes 11: 119–131

25. Diener TO, Hammond RW, Black T, Katze MG (1993) Mechanism of viroid pathogenesis: differential activation of the interferon-induced, double-stranded RNA-activated, 68,000 M_r protein kinase by viroid strains of varying pathogenicity. Biochimie 75: 533–538

26. Diener TO, Lawson RH (1973) Chrysanthemum stunt: a viroid disease. Virology 51: 94–101

27. Diener TO, Raymer WB (1967) Potato spindle tuber virus: A plant virus with properties of a free nucleic acid. Science 158: 378–381

28. Diener TO, Smith DR (1975) Potato spindle tuber viroid. XIII. Inhibition of replication by actinomycin D. Virology 63: 421–427

29. Dinter-Gottlieb G (1986) Viroids and virusoids are related to group I introns. Proc Natl Acad Sci USA 83: 6 250–6 254

30. Domdey H, Sänger HL, Gross HJ (1978) Studies on the primary and secondary structure of potato spindle tuber viroid: products of digestion with ribonuclease A and ribonuclease T_1, and modification with bisulfite. Nucleic Acids Res 5: 1 221–1 236

31. Elena SF, Dopazo J, Flores R, Diener TO, Moya A (1991) Phylogeny of viroids, viroidlike satellite RNAs, and the viroidlike domain of hepatitis δ virus RNA. Proc Natl Acad Sci USA 88: 5 631–5 634

32. Epstein LM, Gall JG (1987) Self-cleaving transcripts of satellite DNA from the newt. Cell 48: 535–543

33. Fauquet CM, Martelli GP (1995) Updated ICTV list of names and abbreviations of viruses, viroids, and satellites infecting plants. Arch Virol 140: 393–413

34. Flores R (1984) Is the conformation of viroids involved in their pathogenicity? J Theor Biol 108: 519–527

35. Fraenkel-Conrat H (1956) The role of the nucleic acid in the reconstitution of active tobacco mosaic virus. J Am Chem Soc 78: 882–883

36. Francki RIB (1987) Encapsidated viroidlike RNA. In: Diener TO (ed) The viroids. Plenum, New York, pp 205–218

37. Gierer A, Schramm G (1956) Infectivity of ribonucleic acid from tobacco mosaic virus. Nature 177: 702–703

38. Gillings MR, Broadbent P, Gollnow BI (1991) Viroids in Australian citrus: relationship to exocortis, cachexia and citrus dwarfing. Aust J Plant Physiol 18: 559–570

39. Grill LK, Semancik JS (1978) RNA sequences complementary to citrus exocortis viroid in nucleic acid preparations from infected *Gynura aurantiaca*. Proc Natl Acad Sci USA 75: 896–900

40. Gross HJ, Domdey H, Lossow C, Jank P, Raba M, Alberty, H, Sänger HL(1978) Nucleotide sequence and secondary structure of potato spindle tuber viroid. Nature 273: 203–208

218   T. O. Diener

41. Gross HJ, Domdey H, Sänger HL (1977) Comparative oligonucleotide fingerprints of three plant viroids. Nucleic Acids Res 4: 2 021–2 028
42. Haas B, Klanner A, Ramm K, Sänger HL (1988) The 7S RNA from tomato leaf tissue resembles a signal recognition particle RNA and exhibits a remarkable sequence complementarity to viroids. EMBO J 7: 4 063–4 074
43. Hadidi A (1986) Relationship of viroids and certain other plant pathogenic nucleic acids to group I and II introns. Plant Mol Biol 7: 129–142
44. Hadziyannis SJ, Taylor JM, Bonino F (eds) (1962) Hepatitis delta virus. Molecular biology, pathogenesis, and clinical aspects. Wiley-Liss, New York
45. Hall TC, Wepprich RK, Davies JW, Weathers LG, Semanik JS (1974) Functional distinctions between the ribonucleic acids from citrus exocortis viroid and plant viruses: Cell-free translation and aminoacylation reaction. Virology 61: 486–492
46. Hammond RW (1995) Viroids. In: Loebenstein G, Lawson RH, Brunt AA (eds) Virus and virus-like diseases of bulb and flower crops. Wiley, Chichester, pp 67–75
47. Hammond RW, Owens RA, Sano T, Diener TO (1995) The role of structural domains in the regulation of viroid pathogenesis. In: Bills DD, Kung S-D (eds) Biotechnology and plant protection: viral pathogenesis and disease resistance. World Scientific, Singapore, pp 201–216
48. Henco K, Riesner D, Sänger HL (1977) Conformation of viroids. Nucleic Acids Res 4: 177–194
49. Hernández C, Elena SF, Moya A, Flores R (1992) Pear blister canker viroid is a member of the apple scar skin subgroup (apscaviroids) and also has sequence homology with viroids from other subgroups. J Gen Virol 73: 2 503–2 507
50. Hernández C, Flores R (1992) Plus and minus RNAs of peach latent mosaic viroid self-cleave *in vitro* via hummerhead structures. Proc Natl Acad Sci USA 89: 3 711–3 715
51. Hiddinga HJ, Crum CJ, Hu J, Roth DA (1988) Viroid-induced phosphorylation of a host protein related to a dsRNA-dependent protein kinase. Science 241: 451–453
52. Hovanessian AG (1989) The double-stranded RNA-activated protein kinase induced by interferon: dsRNA-PK. Interferon Res 9: 641–647
53. Hutchins CJ, Rathjen PD, Forster AC, Symons RH (1986) Selfcleavage of plus and minus RNA transcripts of avocado sunblotch viroid. Nucleic Acids Res 14: 3 627–3 640
54. Jagus R, Anderson W, Safer B (1981) The regulation of initiation of mammalian protein synthesis. Progr Nucleic Acids Res 25: 127–185
55. Katze MG (1992) The war against the interferon-induced dsRNA-activated protein kinase: Can viruses win? J Interferon Res 12: 241–248
56. Katze MG, Tomita J, Black T, Krug RM, Safer B, Hovanessian AG (1988) Influenza virus regulates protein synthesis during infection by repressing the autophosphorylation and activity of cellular 68,000 M_r protein kinase. J Virol 62: 3 710–3 717
57. Keese P, Symons RH (1985) Domains in viroids: Evidence of intermolecular RNA rearrangements and their contribution to viroid evolution. Proc Natl Acad Sci USA 82: 4 582–4 586
58. Klaff P, Gruner R, Hecker R, Sättler A, Theissen G, Riesner D (1989) Reconstituted and cellular viroid-protein complexes. J Gen Virol 70: 2 257–2 270
59. Langowski J, Henco K, Riesner D, Sänger HL (1978) Common structural features of different viroids: Serial arrangement of double helical sections and internal loops. Nucleic Acids Res 5: 1 589–1 610
60. Lawson RH (1968) Some properties of chrysanthemun stunt virus. Phytopathology 58: 885
61. Martínez-Soriano JP, Galindo-Alonso J, Maroon CJM, Yucel I, Smith DR, Diener TO

(1996) Mexican papita viroid: putative ancestor of crop viroids. Proc Natl Acad Sci USA 93: 9 397–9 401

62. McClements WL, Kaesberg P (1977) Size and secondary structure of potato spindle tuber viroid. Virology 76: 477–484

63. Mühlbach H-P, Sänger HL (1979) Viroid replication is inhibited by alpha-amanitin. Nature 278: 185–188

64. Navarro B, Flores R (1998) Chrysanthemum chlorotic mottle viroid: unusual structural properties of a subgroup of viroids with hammerhead ribozymes. Proc Natl Acad Sci USA 94: 11 262–11 267

65. Owens RA, Cress DE (1980) Molecular cloning and characterization of potato spindle tuber viroid cDNA sequences. Proc Natl Acad Sci USA 77: 5 302–5 306

66. Owens RA, Erbe E, Hadidi A, Steere RL, Diener TO (1977) Separation and infectivity of circular and linear forms of potato spindle tuber viroid. Proc Natl Acad Sci USA 74: 3 859–3 863

67. Puchta H, Ramm K, Luckinger R, Hadas R, Bar-Joseph M, Sänger HL (1991) Primary and secondary structure of citrus viroid IV (CVd IV), a new chimeric viroid present in dwarfed grapefruit in Israel. Nucleic Acids Res 19: 6 640

68. Rakowski AG, Szychowski JA, Avena ZS, Semancik JS (1994) Nucleotide sequence and structural features of the group III citrus viroids. J Gen Virol 75: 3 581–3 584

69. Raymer WB, O'Brien MJ (1962) Transmission of potato spindle tuber virus to tomato. Am Potato J 39: 401–408

70. Riesner D (1990) Structure of viroids and their replication intermediates. Are thermodynamic domains also functional domains? Semin Virol 1: 83–99

71. Riesner D (1991) Viroids: From thermodynamics to cellular structure and function. Mol Plant-Microbe Interact 4: 122–131

72. Samuel CE (1991) Antiviral actions of interferon-regulated cellular proteins and their surprisingly selective antiviral activities. Virology 183: 1–11

73. Saville BJ, Collins RA (1990) A site-specific self-cleavage reaction performed by a novel RNA in Neurospora mitochondria. Cell 61: 685–696

74. Sänger HL (1972) An infectious and replicating RNA of low molecular weight: the agent of the exocortis disease of Citrus. Adv Biosci 8: 103–116

75. Sänger HL, Brandenburg E (1961) Über die Gewinnung von infektiösem Pressaft aus 'Wintertyp'-Pflanzen des Tabak-Rattle-Virus durch Phenolextraktion. Naturwissenschaften 48: 391

76. Sänger HL, Ramm (1975) Radioactive labelling of viroid-RNA. In: Modification of the information content of plant cells. North Holland/American Elsevier, Amsterdam, pp 229–252

77. Sänger HL, Klotz G, Riesner D, Gross HJ, Kleinschmidt AK (1976) Viroids are single-stranded covalently closed circular RNA molecules existing as highly base-paired rod-like structures. Proc Natl Acad Sci USA 73: 3 852–3 856

78. Schindler I-M, Mühlbach H-P (1992) Involvement of nuclear DNA-dependent RNA polymerases in potato spindle tuber viroid replication: a reevaluation. Plant Sci 84: 221–229

79. Semancik JS, Morris TJ, Weathers LG, Rodorf BF, Kearns DR (1975) Physical properties of minmal infectious RNA (viroid) associated with the exocortis disease. Virology 63: 160–167

80. Semancik JS, Weathers LG (1968) Exocortis virus of citrus: association of infectivity with nucleic acid preparations. Virology 36: 326–328

81. Semancik JS, Weathers LG (1970) Properties of the infectious forms of exocortis virus of citrus. Phytopathology 60: 732–736

82. Semancik JS, Weathers LG (1972) Exocortis virus: an infectious free-nucleic acid plant virus with unusual properties. Virology 47: 456–466

83. Sharp PA (1985) On the origin of RNA splicing and introns. Cell 42: 397–400

84. Siegel A, Zaitlin M, Sehgal OP (1962) The isolation of defective tobacco mosaic virus strains. Proc Natl Acad Sci USA 48: 1 845–1 851

85. Sogo JM, Koller T, Diener TO (1973) Potato spindle tuber viroid. X. Visualization and size determination by electron microscopy. Virology 55: 70-80

86. Tabler M, Tzortzakaki S, Tsagris M (1992) Processing of linear longer-than-unit-length potato spindle tuber viroid RNAs into infectious monomeric circular molecules by a G-specific endoribonuclease. Virology 190: 746–753

87. Wang K-S, Choo Q-L, Weiner AJ, Ou J-H, Najarian RC, Thayer RM, Mullenbach GT, Denniston KJ, Gerin JL, Houghton M (1986) Structure, sequence and expression of the hepatitis delta (δ) viral genome. Nature 323: 508–514

88. Wolff P, Gilz R, Schumacher J, Riesner D (1985) Complexes of viroids with histones and other proteins. Nucleic Acids Res 13: 355–367

89. Zhang S, Diener TO (1998) A dsRNA-activated protein kinase (PKR) from tomato specifically binds to the left half of the potato spindle tuber viroid. In: Plant viroids and viroid-like satellite RNAs of plants. Instituto Juan March, Madrid, Spain

Author's address: Dr. T. O. Diener, Center for Agricultural Biotechnology, Plant Sciences Building, University of Maryland, College Park, MD 20742, U.S.A.

SpringerLifeSciences

Philip S. Mellor, Matthew Baylis,
Christopher Hamblin, Charles H. Calisher,
Peter P. C. Mertens (eds.)

African Horse Sickness

1998. VIII, 342 pages. 86 partly coloured figures.
Hardcover DM 290,–, öS 2030,–
(recommended retail price)
Special edition of "Archives of Virology, Supplement 14, 1998"
(Softcover edition only available for subscribers to "Archives of Virology")
ISBN 3-211-83133-9

African horse sickness virus is a double-stranded RNA virus which causes a non-contagious, infectious arthropod-borne disease of equines and occasionally dogs. Nine distinct, internationally recognised serotypes of the virus have so far been identified.

This book is based upon the findings of two programmes funded by the European Commission. It will be of value not only to the specialist research workers but also to veterinary workers dealing with control and to legislators seeking to promote safe international movement of equines.

The topics covered include state-of-the-art discussions on diagnostics, vaccines, molecular biology, vector studies, and epidemiology.

Contents
- Epidemiology
- Entomology
- Molecular Biology
- Vaccines and Diagnosis

SpringerWienNewYork

Sachsenplatz 4–6, P.O.Box 89, A-1201 Wien, Fax +43-1-330 24 26, e-mail: books@springer.at, **Internet: http://www.springer.at**
New York, NY 10010, 175 Fifth Avenue • D-14197 Berlin, Heidelberger Platz 3 • Tokyo 113, 3–13, Hongo 3-chome, Bunkyo-ku

SpringerVirology

Oskar-Rüger Kaaden,
Claus-Peter Czerny,
Werner Eichhorn (eds.)

Viral Zoonoses and Food of Animal Origin

A Re-Evaluation of Possible Hazards for Human Health

1997. VIII, 256 pages. 43 figures.
Hardcover DM 248,–, öS 1736,–
(recommended retail price)
Special edition of Archives of Virology, Supplement 13, 1997
(Softcover edition only available for subscribers to "Archives of Virology")
ISBN 3-211-82927-X

The investigations of virus infections naturally transmitted from animals to men are a challenge to multidisciplinary science. Some of these zoonoses are very common, others are sporadic but show a life-threatening clinical course.

The contributions presented by world-wide leading experts are going to update the present scientific, administrative and legislative knowledge in the field of food-borne virus infections in men. Major topics include classic zoonoses, pox, irido, influenza, enteric and newly emerging virus diseases and their role for Public Health including strategies to avoid virus transmission by biopharmaceutical products.

The contributions are aimed at public and veterinary public health authorities, diagnostic and scientific institutes as well as food producing and pharmaceutical companies.

SpringerWienNewYork

Sachsenplatz 4–6, P.O.Box 89, A-1201 Wien, Fax +43-1-330 24 26, e-mail: books@springer.at, **Internet: http://www.springer.at**
New York, NY 10010, 175 Fifth Avenue • D-14197 Berlin, Heidelberger Platz 3 • Tokyo 113, 3–13, Hongo 3-chome, Bunkyo-ku